Active Vibration Control and Stability Analysis of Flexible Beam Systems

Active Vibration Control and Stability Analysis of Flexible Beam Systems

Wei He · Jinkun Liu

Active Vibration Control and Stability Analysis of Flexible Beam Systems

清華大學出版社
TSINGHUA UNIVERSITY PRESS

Springer

Wei He
School of Automation and Electrical Engineering
University of Science and Technology Beijing
Beijing, China

Jinkun Liu
School of Automation Science and Electrical Engineering
Beihang University
Beijing, China

ISBN 978-981-10-7538-4 ISBN 978-981-10-7539-1 (eBook)
https://doi.org/10.1007/978-981-10-7539-1

Jointly Published with Tsinghua University Press, Beijing, China

The print edition is not for sale in China Mainland. Customers from China Mainland please order the print book from: Tsinghua University Press.

Library of Congress Control Number: 2018953307

This Springer imprint is published by the registered company Springer Nature Singapore Pte Ltd.
The registered company address is: 152 Beach Road, #21-01/04 Gateway East, Singapore 189721, Singapore

Preface

Control of flexible beam systems constitutes an important benchmark problem in many application areas, such as flexible manipulators for grasping, marine mooring lines for station keeping, marine risers for oil transportation, and crane cables for positioning the payload. Flexible systems have a number of advantages such as lightweight, efficiency, higher operation speed, and lower cost. The dynamics of the flexible beam system is essentially a distributed parameter system (DPS). Different from lumped parameter systems, the DPS has an infinite-dimensional state space. The dynamics of DPS modeled by the coupled partial differential equations (PDEs)–ordinary differential equations (ODEs) is difficult to control due to the infinite dimensionality of the beam system.

Since the flexible beam system becomes lighter and more flexible, the external disturbances will lead to the mechanical vibrations of flexible beam systems. Therefore, the control strategy should be designed to suppress the vibrations of flexible beam systems. Driven by practical needs and theoretical challenges, flexible beam systems and their vibration suppression have received great attention. Boundary control is an effective control strategy for vibration suppression of the flexible beam system which is described by hybrid PDEs–ODEs. By using sensors and actuators at the boundary, the dynamic model of the system is not affected, and boundary control can be derived from a Lyapunov function which is relevant to the mechanical energy based on the dynamics of the system.

The purpose of this book is to investigate the fundamental issues including dynamic analysis and control design for flexible beam systems by theoretical analysis and numerical simulations. A comprehensive study is provided to develop boundary control methods for the vibration suppression of flexible beam systems with input nonlinearities and output constraint. In addition, the book presents theoretical explorations for advanced control methods of flexible beam systems, including distributed control, iterative learning control, and neural network control. The control designs are coupled with numerical simulations to illustrate the effectiveness.

The book starts with a brief introduction of modeling methods and control techniques for a class of flexible beam systems in Chap. 1.

Chapter 2 presents the preliminaries and several lemmas for the subsequent development to simplify the dynamical modeling and further stability analysis for the beam structures, and the dynamic model is also introduced.

In Chap. 3, we investigate the boundary control design problem for an Euler–Bernoulli beam model under both the unknown spatiotemporally varying distributed disturbance and time-varying boundary disturbance. The adaptive control laws and disturbance observers are proposed to deal with system parametric uncertainties and external disturbances, respectively.

In Chap. 4, the control problem of an Euler–Bernoulli beam with boundary output constraint is addressed. We design a boundary barrier control scheme and apply the proposed Lyapunov function to the original partial differential equations in order to avoid the spillover problem. The form of Lyapunov function combining both the integral Lyapunov function and the barrier Lyapunov function is employed for the control design and stability analysis of the system.

In Chap. 5, the largest challenge is how to eliminate the nonlinear input saturation characteristic and to design an effective active control law for the flexible beam system. We propose the hyperbolic tangent function in the control design of the flexible system to reduce the influence induced by the input saturation.

In Chap. 6, the boundary control architecture is proposed for continuous-time PDE systems preceded by input dead-zone nonlinearity. By using a new description of the dead-zone, boundary control schemes are developed to regulate the deformation of Euler–Bernoulli beam system even in the presence of the external disturbance. It is proved that the proposed control law can ensure uniformly ultimately boundness of the entire system and achieve stabilization.

In Chap. 7, we discuss a problem of a flexible beam system with input backlash characteristic. By transforming the nonlinear input backlash to linear input, the boundary control law and disturbance observer are designed at the top of flexible beam via constructing a Lyapunov candidate function to reduce the vibration of the system. The transverse displacement of the closed-loop system is proved to converge to a small neighborhood of zero.

In Chap. 8, we present the control design for the flexible Euler–Bernoulli beam system with global constraint and uncertain tip payload. To prevent the constraint violation, the integral barrier Lyapunov function is employed for the control design and stability analysis. We propose a distributed control so that the deflection of the mechanical system can track the desired signal. Exponential stability is well achieved without violation of the constraint.

In Chap. 9, we propose an adaptive boundary iterative learning control (ABILC) scheme for an Euler–Bernoulli beam system with an aperiodic distributed disturbance, an aperiodic boundary disturbance, and an unknown system parameter. In order to tackle the input saturation, a hyperbolic tangent function and a saturation function are utilized. Three adaptive laws are proposed and learned along the iteration axis. Based on a time-weighted Lyapunov–Krasovskii-like composite energy function, the restrained ABILC law is designed. For the closed-loop system, the boundedness of all the signals in each iteration is guaranteed. Along the iteration axis, the displacements converge to zero in the presence of the external disturbances.

In Chap. 10, in order to analyze the flexible beam structure, the assumed mode method (AMM) is employed to develop the dynamic model. Based on the N-dimensional discrete dynamic model, neural network control is investigated to track the desired trajectory accurately and to suppress the flexible vibration maximally. To ensure stability rigorously as the goal, the system is proved to be uniform ultimate boundedness (UUB) by Lyapunov stability method. Eventually, simulations verify that the proposed control strategy is effective.

In Chap. 11, a model of a coupled three-dimensional flexible beam with a tip payload is derived by using Hamilton principle and described by a set of partial differential equations and ordinary differential equations. Boundary control laws are proposed to suppress the vibration of the beam in both longitudinal and transverse directions with the environmental disturbance, and the uniform boundedness and the uniformly ultimate boundedness of the closed-loop system are proved. Adaptive control laws are proposed to overcome the uncertainty of the tip payload, and the uniform boundedness and the uniformly ultimate boundedness of the system are proved.

In Chap. 12, the conclusions are summarized and the future works are given.

In summary, this book covers the dynamical analysis and control design for flexible beam systems. The book is primarily intended for researchers and engineers in the control system. It can also serve as a complementary reading on modeling and control of flexible beam systems at the postgraduate level.

Beijing, China

Wei He
Jinkun Liu

Acknowledgements

For this book, we have had the great fortune of working with brilliant people who are generous with their time and friendship, through many discussions filled with creativity and inspiration. First of all, we would like to express our gratitude to our co-workers who have contributed to the collaborative studies of this book.

We would also like to express our sincere appreciation to our colleagues who have contributed to the collaborative research. In particular, we would like to thank Shuzhi Sam Ge, from the National University of Singapore, Singapore; Miroslav Krstic, from the University of California, San Diego, US; Khac Duc Do, from the Curtin University, Australia; Keum-Shik Hong, from the Pusan National University, Korea; Bao-Zhu Guo, from the Academy of Mathematics and Systems Science, Chinese Academy of Sciences, China; Huai-Ning Wu, from the Beihang University, China; Jun-Min Wang, from the Beijing Institute of Technology, China, and their research groups for their excellent research works, and helpful advice on our research. Special thanks go to Dongliang Sheng for his assistance and efforts on the process of publishing this book.

Appreciations must be made to Xiuyu He, Yuhua Song, Tingting Meng, Kai Huang, Hejia Gao, Zhe Jing, Weijie Xiang, Hui Qin, Hoang Minh Vu, Yu Liu, Zhijia Zhao, Zhijie Liu, Yuncheng Ouyang, Linghuan Kong, Xinling Yue, Jiali Feng, and Qingyu Zhou for the constructive discussions and sharing of ideas.

This work is supported by the National Key Research and Development Program of China under Grant 2017YFB1300102, the National Natural Science Foundation of China under Grant 61522302, 61873298, 61761130080, the Newton Advanced Fellowship from The Royal Society, UK, under Grant NA160436, the Beijing Natural Science Foundation under Grant 4172041, and the Fundamental Research Funds for the China Central Universities of USTB under Grant FRF-BD-17-002A.

Haidian District, Beijing, China

Wei He
Jinkun Liu

Contents

Chapter 1
Introduction

1.1 Background and Motivation

Over the past years, the flexible beam structure is widely used in modern engineering because of its advantages (e.g. light weight, low energy consumption, etc.), and its control problem becomes one of the hot research topics [1, 2]. In the traditional rigid structure, the positional relationship between adjacent particles is constant. However, each particle of the flexible beam structure is relatively independent varying. Figure 1.1 shows the deformation of a typical flexible beam structure in the presence of external disturbances.

For the space manipulator, the vibrating arm will affect the precision operation, even lead to fatigue damage and shorten the lifespan. The slow-reducing vibration will have great influence on the flexible system because of no atmospheric damping. In the field of mechanical engineering, with the development of robotics technology with high speed, high precision and lightweight, industrial robots are also changing from rigid arms to flexible arms. The small damping characteristics of flexible arms affect the positioning accuracy and motion stability of industrial robots. In the field of offshore oil and gas development, the deep-sea oil pipeline can be regarded as a flexible beam structure. If the flexible beam system cannot be well controlled, the vibration will not only affect the accuracy and efficiency of the system, but also accelerate the equipment fatigue damage, seriously shorten the service life of the materials, and bring production safety risk and economic loss. Therefore, it is very important to effectively control flexible beam systems. Driven by practical needs and theoretical challenges, flexible beam systems and their vibration suppression have received great attention in the literatures [3–5]. Boundary control [6–8] has several merits for vibration suppression of the flexible beam systems. The distributed control is not practical to be implemented and the distributed system becomes uncontrollable and unobservable when the point actuators and sensors are located at nodal points [9]. For boundary control, by using sensors and actuators at the boundary, the dynamic model of the system is not affected, and boundary control can be derived from a Lyapunov function which is relevant to the mechanical energy based on the dynamics

W. He and J. Liu, *Active Vibration Control and Stability Analysis of Flexible Beam Systems*, https://doi.org/10.1007/978-981-10-7539-1_1

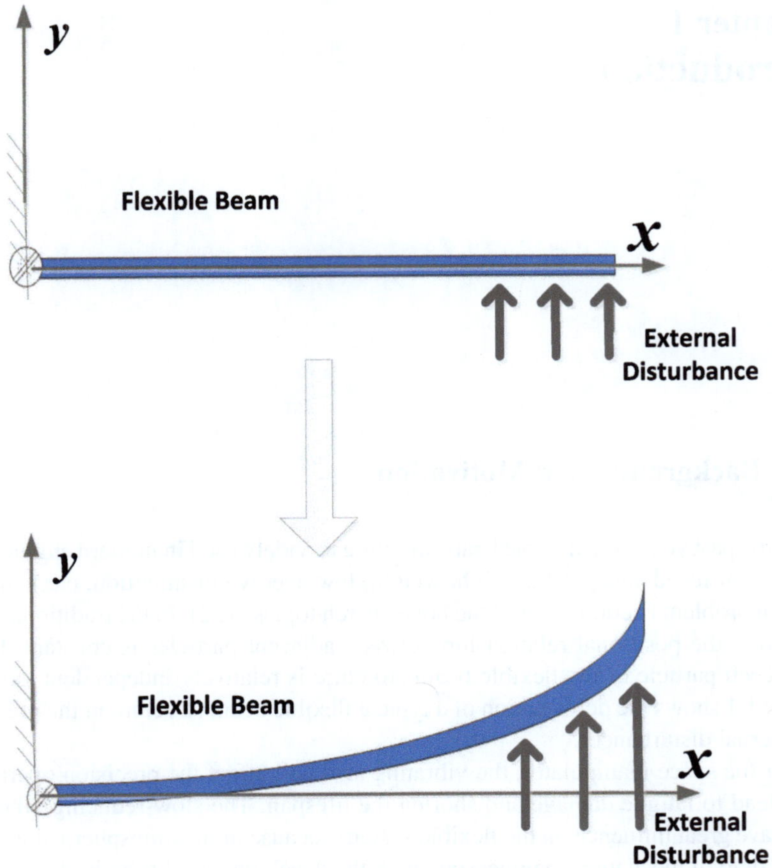

Fig. 1.1 Vibration diagram of a flexible beam

of the system [10, 11]. Due to these advantages, boundary control has been employed in a number of research fields. In [12], by using a positive real (PR) transfer function, the transverse vibration of a string system is controlled by boundary hydraulic touch rolls. The authors in [13] propose two boundary control laws to regulate the vibrations for a Timoshenko beam modeled by a coupled non-linear system. The control theory and experimental verification are discussed in [14] for an axially moving string system consisting of a controlled span coupled to a disturbed span via an actuator. In [15, 16], the iterative learning control (ILC) scheme is developed for boundary control of a class of PDEs. In [17], stabilization of linearized 2D magnetohydrodynamic channel flow is studied by using the backstepping boundary control, where the local stability is guaranteed. Recently, by using the backstepping method, a novel boundary controller/observer is designed to stabilize the string and beam model and to track the target system [18, 19].

Lyapunov theory, the most successfully and widely used tool, provides a means of determining stability without explicit knowledge of system solutions [20, 21].

Control design and stability analysis based on the Lyapunov approaches for a number of nonlinear systems have got significant results these years. In [22], stability analysis is performed for a class of nonlinear systems which are described by discrete Takagi and Sugeno (TS) fuzzy models by using new control laws and new nonquadratic Lyapunov functions. In [23], under the SOS-based framework, the output-feedback tracking control problem for polynomial fuzzy-model-based control systems is studied on the basis of the Lyapunov stability theory. In [24], two indirect adaptive fuzzy control schemes are discussed for a class of uncertain continuous-time multi-input-multi-output (MIMO) nonlinear dynamic systems based on Lyapunov stability analysis.

The flexible beam is a common structure in industrial production, and its control problem has reached quite rich theoretical results, but there is no uniform and universal conclusion. The flexible beam structure is a kind of typical infinite dimensional distributed parameter systems. The dynamics of this kind of systems are represented by infinite modal, and they have nonlinearity, strong coupling and time-varying characteristics, which make the research of flexible beam systems quite difficult. The difficulty lies in the establishment of the system dynamics model and the choice of control strategies. The following sections describe the modeling and control of flexible beams, respectively.

1.2 Application of Flexible Beam Systems

In the actual environment, flexible beam systems have universal applications: the marine riser system [25, 26], the flexible manipulator system [27, 28], the flexible moving system [29], the flapping-wing robotic aircraft system, the fire rescue turntable ladder system, the wind turbine tower system [30, 31], etc.

In the ocean, the marine riser shown in Fig. 1.2 connects a production vessel and a subsea oil wellhead. The drilling riser [32], as a core component of the drilling system, was generally considered as a flexible beam system. Reference [8] considered the flexible riser system with distributed vibration and used the Lyapunov's direct method to design the control law. Reference [33] proposed a model of extensible marine risers in three-dimensional space under sea loads and designed boundary control for practical exponential stabilization. The riser installation system which was proposed in [13] includes a vessel, a flexible riser, and a subsea payload, and the system was described by partial differential equations and ordinary differential equations.

The flexible manipulator is regarded as a rotatory flexible beam system. Figure 1.3 shows a single-link flexible manipulator, and Fig. 1.4 shows a two-link flexible manipulator. To suppress the vibration, the neural network control method was applied to the flexible robotic manipulator system. The integral resonant control scheme was used to achieve the suppression of vibration and positioning of the tip position for a single-link flexible manipulator [34]. In [35], the authors presented the partial differential equation model to describe a flexible two-link manipulator system, and designed

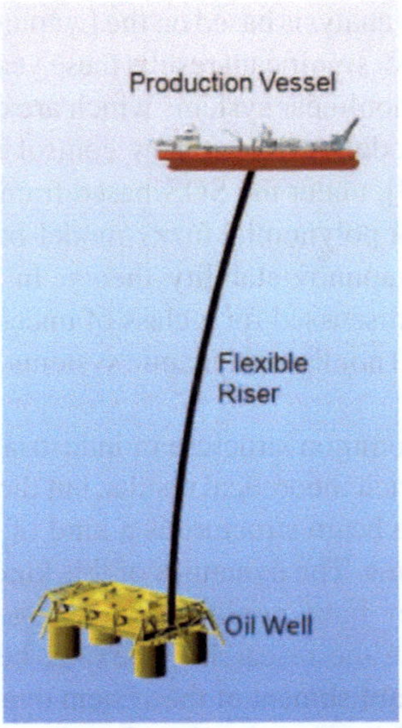

Fig. 1.2 A marine riser system

Fig. 1.3 A single-link flexible manipulator system

the control method based on the Lyapunov's direct method. A flexible manipulator which had different payloads was proposed in [36], and the nonlinear adaptive model predictive controller was used to handle the tip position control problem.

The flexible moving system, like moving belts, chains, etc., is very common in industrial production. The surface mount technology equipment in Fig. 1.5 is an application of the moving belt system. In [37], adaptive boundary control was proposed for an axially moving belt system to suppress the vibration. Reference [38] considered an active pivoting roller system as a flexible beam structure, and used an adaptive vibration isolation method to isolate a controlled span from bounded disturbances. The translating steel strip was also considered as the moving belt system

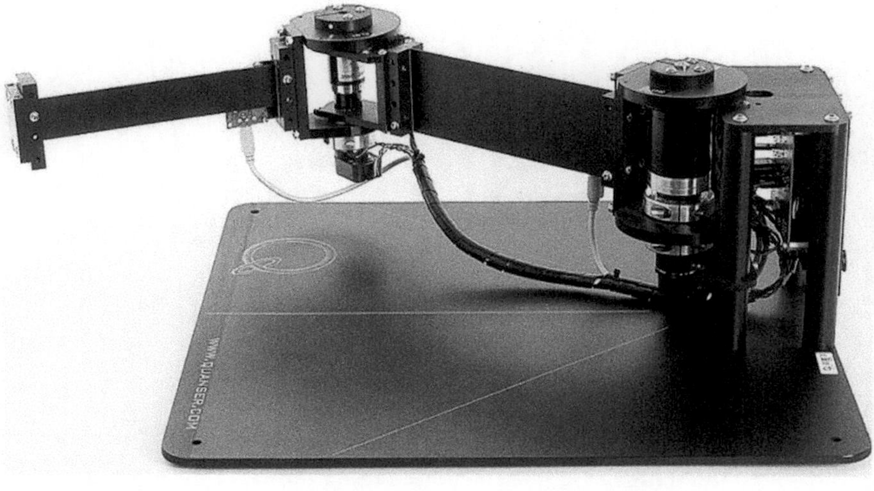

Fig. 1.4 A two-link flexible manipulator system

Fig. 1.5 The surface mount technology equipment

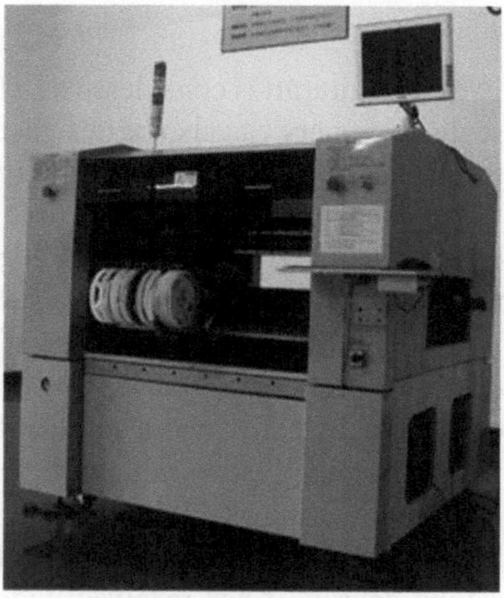

in [39], and boundary control was designed to improve the uniformity of steel strip surface.

Flexible wings of the robotic aircraft are widely used in micro aerial vehicles [40] which have gradually become a hot issue in aerospace engineering research. As shown in Fig. 1.6, it is a flapping-wing robotic aircraft named Smartbird which uses flexible wings. Reference [41] studied flexible wings of a robotic aircraft. It used the partial differential equations and ordinary differential equations to model, and imposed the control force at the boundary. The small robotic aircraft was designed with articulated flexible wings [42], and its output is the net aerodynamic force or moment. Authors in [43] considered aeroelastic wing micro-autonomous aerial systems as two flexible beams, and dealt with the sensitivities and functional gain problems.

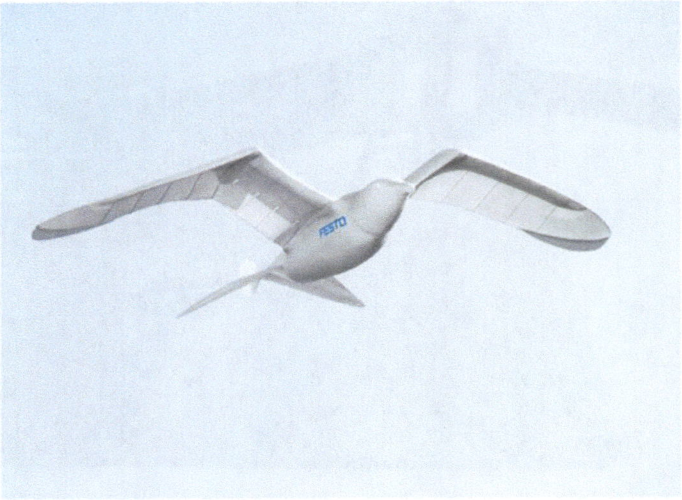

Fig. 1.6 Smartbird of Festo

Micro-aircraft is considered as a flexible beam structure, on the other hand, large-scale machinery can also be designed as a flexible beam model to control. In Fig. 1.7, the fire rescue turntable ladder is a typical large scale robot. The fire-rescue turntable ladder can be modeled as a flexible beam structure, and model parameters are discontinuous [44]. Then, the two degrees of freedom control can be used to ensure that the system is asymptotically stable. Reference [45] studied the articulated aerial work platform, which is a kind of large-scale vehicles with a flexible beam. Its control objectives were trajectory tracking and vibration suppression. The fire rescue turntable ladder was also regarded as a piecewise flexible beam system in [46], which discussed the vibration of the truss structure in a vertical plane.

Figure 1.8 shows a wind turbine tower which exploits wind power (a renewable resource) to replace fossil fuels for power generation. Because of advantages of saving energy, it has received widespread attention. The wind turbine tower was modeled as a flexible beam system, which was affected by unknown distributed disturbances generated by random winds [47]. Then the control force was applied at the top to suppress the vibration. Reference [48] considered the wind turbine tower as an elastic Euler-Bernoulli beam, and used the separation of a variable method to manage the model. In [49], the parameter analysis and modeling of a passive structural control strategy for the OC3-Hywind floating wind turbine system were discussed.

These examples illustrate that applications of flexible beam systems in practical production are very common, therefore the study of the flexible beam is necessary and meaningful. This book focuses on the system modeling and control design of flexible beams. The structure of this chapter is as follows: Section 1.3 outlines types

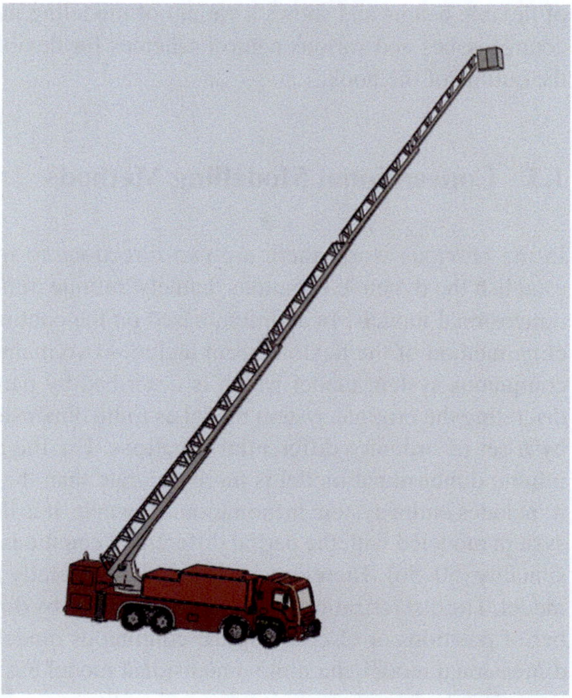

Fig. 1.7 The fire rescue turntable ladder

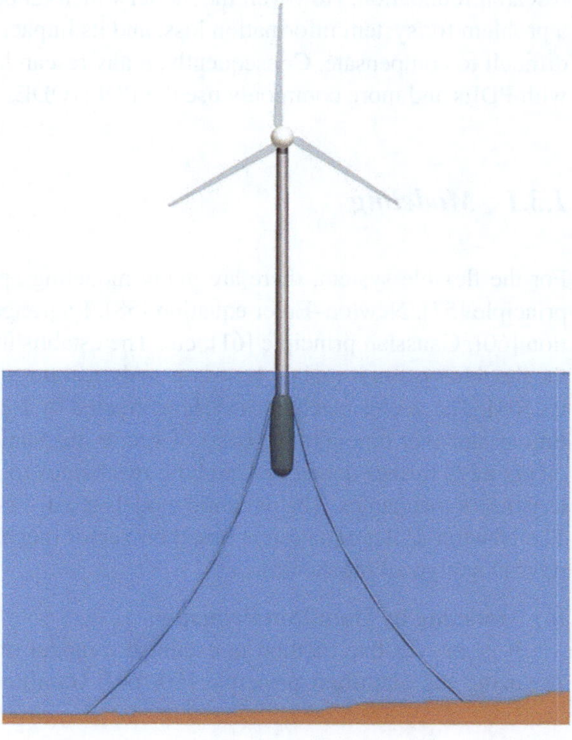

Fig. 1.8 The floating wind turbine towers

of flexible beams and shows a variety of modeling methods. Section 1.4 introduces control issues and various control schemes for flexible beams. Section 1.5 presents the outline of the book.

1.3 Conventional Modelling Methods

In the previous work, there are two directions to model the flexible system and establish the dynamic equations, namely, infinite dimensional models and the finite dimensional models. In addition, based on the continuous system model, the modeling method of the flexible beam includes two main categories: using the original continuous system model which is described by partial differential equations and discreting the original system model as finite dimensional model which is described by a set of ordinary differential equations. For the actual system, the continuous infinite dimensional model is more accurate than the finite dimensional model and it includes entire system information. However, it is difficult to solve for the flexible system modeled with the partial differential equations because of the infinite dimensionality [50–56]. Therefore, earlier studies generally focus on the finite dimensional model. The discretization model can be obtained by directly considering a finite number of positions or discretizing the continuous model. Comparing with the infinite dimensional model, the finite dimensional model has some advantages, e.g., simple structure characteristic, finite dimensionality, easy mathematic model and mature research foundation. However, the model which ignores some modes may generate a problem for system information loss, and its impact on the control performance is difficult to compensate. Consequently, many researches are increasingly concerned with PDEs and more commonly use the PDEs-ODEs model.

1.3.1 Modeling

For the flexible system, there are many modeling approaches including Hamilton principle [57], Newton–Euler equation [58], Lagrangian equation [59], Kane equation [60], Gaussian principle [61], etc. The establishment of the dynamic model of the flexible system is mainly based on analytical mechanics [62] and vector mechanics [63]. The vector mechanics system, created by Newton with vector form, shows limitations over time. On the basis of vector mechanics, Lagrangian, Hamilton and others have further developed analytic mechanics to analyze mathematical characteristics of mechanics. The dynamic models are diverse, including PDEs and ODEs. The Newton–Euler equation is based on vector mechanics, and the rest of methods refer to analytical mechanics.

(a) **Modeling by Hamilton Principle**:
It is an effective method to establish a model of the flexible beam system by using the Hamilton principle [50, 64]. Hamilton principle models a flexible

beam system by considering the energy mode such as kinetic energy, potential energy, virtual work and virtual momentum transport [25], and avoids internal force terms [65, 66]. Hamilton principle is usually written as $\int_{t_1}^{t_2} \delta(E_k - E_p + W)dt = 0$, where E_k is the kinetic energy, E_p is the potential energy, and W is the virtual work done by the nonconservative force [67]. The model obtained by this method satisfies the energy conservation law, namely the above equation is a form of the energy conservation law. This model is a continuous system based on the PDEs and combines density, bending stiffness, tension, length, and other system parameters in the equation to fully describe the system characteristics. This method is complex due to the high system order and infinite dimensionality, however it still gets more and more extensive attention because it can perfectly express system information to achieve precise control.

(b) **Modeling by Newton–Euler equation**:
In the flexible beam modeling based on the Newton–Euler equation, the motion decomposition of the system is the translation based on the Newton equation and the rotation based on the Euler equation [68]. The Newton–Euler equation is the basis of many dynamic algorithms, which applies the centroid momentum theorem to express the complete force relations of the system, however, the dynamic modeling based on the Newton–Euler equation contains more ideal binding force. The Newton–Euler equation is an early common method.

(c) **Modeling by Lagrangian Equation**:
It is general to study the dynamics of flexible beam systems based on the Lagrangian equation [69–71]. Motion equations are written based on the Lagrangian equation, and constraint equations are necessary to write simultaneously. This modeling method based on Cartesian generalized coordinates can obtain the differential-algebraic equation of the system, and the numerical solution of a differential-algebraic equation is an important research direction.

(d) **Modeling by Kane Equation**:
The Kane equation is a general dynamical method for a complicated rigid-flexible coupling multi-body system [72]. The pseudo-velocity is introduced as a generalized coordinate to describe the motion of the system and the kinetic equation is established by D'Alembert's principle. It combines the advantages of analytical mechanics and vector mechanics, and the derivation is normalized. The equation is a first order differential equation group, which is convenient for computer programming.

(e) **Modeling by Gaussian Principle**:
Gaussian principle is also called Gaussian minimum restraint principle, which is one of the general differential variational principles in analytical mechanics [73, 74]. Gaussian principle can be expressed as: the real acceleration of the system of particles is the one which can let the constraint function take a minimum in all the possible accelerations that satisfy the constraint. On the basis of Gauss principle, dynamic modeling analyzes the possible movement of the flexible system according to the variational principle rather than according to the specific principle of the movement, and uses the principle of functional extremum to solve the motion law of the flexible beam system.

1.3.2 *Model Discretization*

The discretization of continuous systems is a common means to obtain ordinary differential equations which have plenty of research results and control schemes. The widely used discretization approaches are assumed mode method (AMM) [75], finite element method (FEM) [76, 77], finite difference method (FDM) [78], and lumped parameter method (LPM) [79], etc.

(a) **Assumed Mode Method**
AMM is a generalized coordinate approximation, and it uses a linear combination of finite known modal functions to approximate the response of the system [80, 81]. In fact, the AMM is a method for discretization of the dynamic equations of flexible beams. An advantage that makes it interesting for a control design is that the model derived with AMM is of low order. However, the decision for a correct mode assumption is not straightforward. It is not possible to use exact modes for the description of the elastic deflection. By increasing the number of assumed modes, more precise dynamic models may be expected, but the model will be more complex for controller designs.

(b) **Finite Element Method**
FEM is a numerical method [82] for solving problems of flexible beam systems. Continuous systems are based on PDEs and the analytical solution of PDEs is generally difficult to obtain. The continuous solution field is transformed into the combination of a set of elements, and the approximate function assumed in each element is used to represent the unknown field function to be solved in the solution domain. Approximate functions are the numerical interpolation functions of the unknown field function and its derivatives at each node of the element. So that a continuous infinite dimensionality system becomes a discrete finite dimensionality system. The basic idea is to solve the extreme value problem of functional by solving the Poisson equation.

(c) **Finite Difference Method**
FDM is another numerical solution [83]. The basic idea is to decompose the problem definition domain into grids first. Then, at the grid point, the differential quotient is changed into the differential coefficient in the problem for determining solution according to the appropriate numerical differential formula. Thus the original problem is dispersed into a differential format, and then find the numerical solution. It is simple, versatile and easy to implement on a computer.

(d) **Lumped Parameter Method**
LPM integrates all the mass of the flexible beam system to a node, and then replaces the distribution quality of the original flexible beam system with the concentrated mass on several discrete nodes. The essence of this method is to use the discretization to approximate the quality of the whole flexible beam system, so as to obtain the dynamic equation.

1.4 A Brief Survey of Control Approaches

Main control issues for flexible beam systems are the following four aspects: position control [84–87], force control [88–90], vibration control [65, 91–93] and hybrid control [94–96]. The goal of position control is to enable a flexible beam to reach a specified position or to track a desired trajectory. The goal of force control is to make the detected force reach a desired requirement. The goal of vibration control is to suppress vibration (generated by its own structure and external disturbance) to ensure the stability of a closed-loop system. Hybrid control is a combination of a variety of control methods to achieve multiple control targets.

For the control of flexible beams, many studies have proposed solutions, which are generally divided into passive control and active control. Passive control [97] is installing a mechanical device that can suppress the vibration of a flexible beam system. The effect of the passive control is obvious, but this control method has drawbacks such as high weight, high cost, etc. Therefore, many scholars begin to study the active control strategy. The active control method [98, 99] analyzes the dynamic model of a system and designs a control algorithm based on the model. The desired damping characteristic is obtained by the input of an external energy, and the elastic vibration of a flexible beam is finally suppressed. Active control can be divided into two methods, namely, open-loop control and closed-loop control. The open loop control considers the flexible structure and realizes the power or torque of the system. This method requires an accurate dynamic model and is sensitive to changes of system parameters and external disturbances. Closed-loop control is based on the feedback of the system state to generate the power or torque, in order to achieve the control of a flexible beam. It has the ability to suppress disturbances and is insensitive to changes in component characteristics, but increases the complexity of the system. Control schemes and their characteristics are listed in Table 1.1.

1.4.1 Implementation Location

The control applied at different positions results in dynamic models with different characteristics and requires different control design strategies. By comparing controlled positions, the control can be divided into three types: boundary control, distributed control and point control.

(1) **Boundary Control**

Boundary control is a method of applying control force at the boundary of the structure [100–102]. Figure 1.9 shows a typical boundary control design for a flexible beam system. When sensors and actuators are applied on the boundary, the dynamic model will not be affected [103, 104]. The design of boundary control is based on a infinite dimensional dynamic model, and does not approximate the model, which can effectively avoid the spillover effect that may be produced by neglecting the high frequency mode. Boundary control requires only a few

Table 1.1 Control schemes and their characteristics

Control schemes	Characteristics
Boundary control	Simple structure, easy implementation, low cost, but difficult to design
Point control	The difficulty depends on the choice of control nodes
Distributed control	A large number of sensors and actuators, difficult to implement
PID control	Error-based, linear controller, simple structure, easy implementation
Model-based control	The accuracy of the model directly affects the control effect
Robust control	Invariant controller, ensure the control effect in different environments
Adaptive control	Variable controller, adapt the uncertain system by online changing control parameters
Optimal control	Optimization algorithm, it is difficult to obtain analytical solution to nonlinear systems, but is easy for linear systems
Sliding mode control	Variable structure, good robustness, simple realization
Iterative learning control	Try the same control, modify the control parameters with the error
Neural network control	Less system information, learning ability, need time to online train and learn
Fuzzy logic control	No precise model is required, difficult to choose parameters

actuators, and this feature has a significant advantage in practical engineering, such as simple structure, easy implementation, low cost and so on [105]. Boundary control is a suitable and practical control method for the accuracy analysis of system and the implement of the controller [106, 107].

(2) **Point Control**

Point control is an extension of boundary control [108, 109] and this control method is to apply the control force at the boundary or at the nodes of the flexible beam shown in Fig. 1.10 [110]. Compared with the boundary control, it is easy to design with the point control strategy because there are several controllers used to the flexible structure. However, its practical application will become more difficult because the number of nodes increases and the location of the nodes is uncertain.

(3) **Distributed Control**

The way of adding more modal by increasing the number of sensors is the core idea of distributed control in order to avoid the spillover effect [111, 112]. In Fig. 1.11, the flexible beam is considered as a distributed parameter system to control. Further, it is unavailable to add more sensors which are installed at the flexible structure to detect the feedback signals for the control design of the flexible beam. Because of the increased number of sensors, it is easy to derive the control algorithm in theory with the distributed control, and the derived algorithm also can get good control effect [113, 114]. However, it will undoubtedly increase the number of actuators and the controller in practical application which will greatly reduce the practicality and become more difficult to implement for this control method [9].

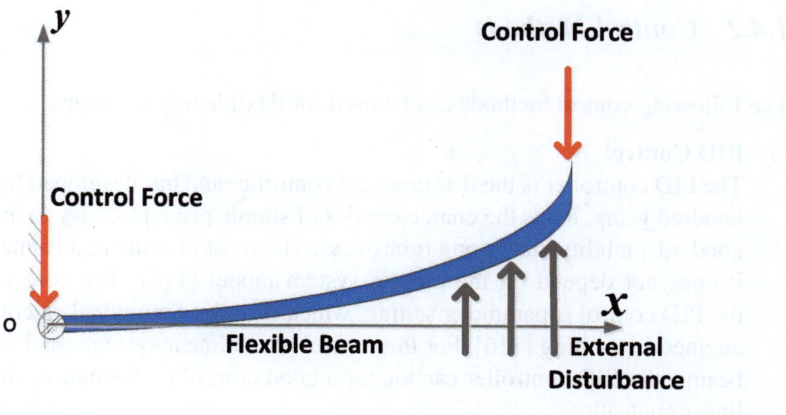

Fig. 1.9 Boundary control

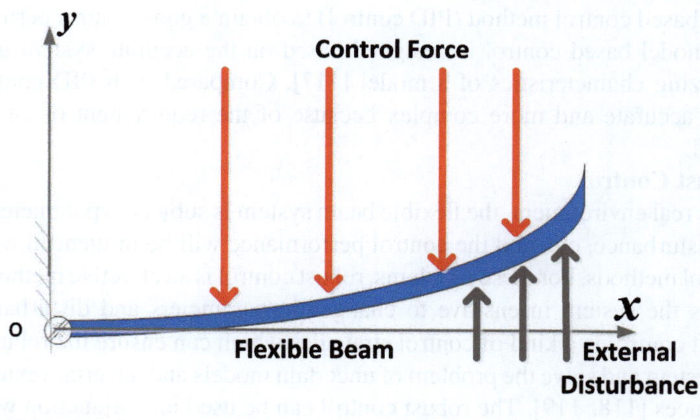

Fig. 1.10 Point control

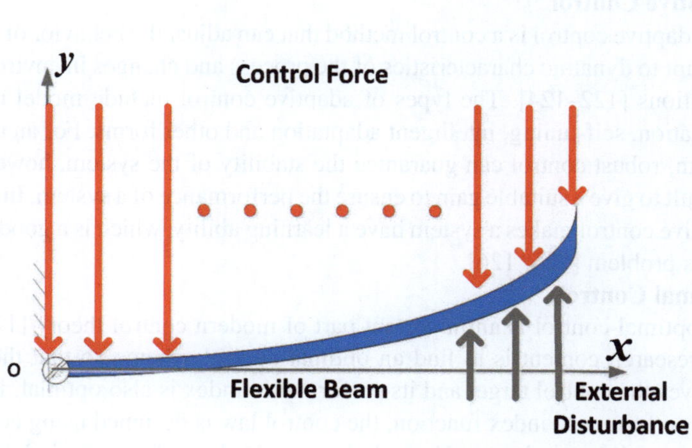

Fig. 1.11 Distributed control

1.4.2 Control Methods

The following control methods can be used for flexible beam systems:

(1) **PID Control**
The PID controller is the first practical controller and has developed for nearly a hundred years. It has the characteristics of simple principle, easy to implement, good adaptability and strong robustness. The most prominent advantage is that it does not depend on the precise system model [115]. The core content of the PID control is parameter setting, which includes theoretical calculation and engineering setting [116]. For the control of nonlinear systems such as flexible beams, the PID controller cannot get a good control performance, since it is a linear controller.

(2) **Model-Based Control**
Because of the complexity of the flexible beam system, it is difficult for the error-based control method (PID control) to obtain a good control performance. The model-based control is designed based on the accurate system model by analyzing characteristics of a model [117]. Compared with PID control, it is more accurate and more complex because of the requirement of an accurate model.

(3) **Robust Control**
In the real environment, the flexible beam system is subject to parameters, external disturbance, etc. and the control performance will be influenced with some control methods. For these problems, robust control is an effective method, which makes the system insensitive to changes in parameters and disturbance. The robust control is a kind of control strategies which can ensure the robustness of the system and solve the problem of uncertain models and uncertain external disturbances [118, 119]. The robust control can be used in conjunction with other control methods (boundary control, PID control, model-based control, etc.) to improve the robustness of a system [120, 121].

(4) **Adaptive Control**
The adaptive control is a control method that can adjust the behavior of a system to adapt to dynamic characteristics of the process and changes in environmental conditions [122–124]. The types of adaptive control include model reference adaptation, self-tuning, intelligent adaptation and other forms. For an uncertain system, robust control can guarantee the stability of the system, however, it is difficult to give a suitable gain to ensure the performance of a system. In contrast, adaptive control makes a system have a learning ability, which is a good solution to this problem [125, 126].

(5) **Optimal Control**
The optimal control is an important part of modern control theory [127, 128]. The research content is to find an optimal control scheme so that the system achieves the control target and its performance index is also optimal. By defining a performance index function, the control law is designed using constraints such as motion equations and boundaries to make the performance index function

obtain the extremum. The main methods to solve the optimal control problems are a classical variational method, maximum principle and dynamic programming. The optimal control problem of the flexible beam and the reduction of the controller have some research results.

(6) **Sliding Mode Control / Variable Structure Control**
The sliding mode control [5, 129], also known as variable structure control or sliding mode variable structure control [130], is a special nonlinear control which can be used in flexible beam systems. According to the current states of the system (such as the deviation and its derivatives, etc.), the output of sliding mode control has continuous change which makes the system track a given mode [131]. The general sliding mode control system is composed of two stages: approaching motion and sliding mode motion. It has the advantages of good robustness and simple realization, and it is widely applied to the fields of nonlinear system and robust control [6].

(7) **Iterative Learning Control**
For nonlinear control systems such as a flexible beam, the iterative learning control has excellent control effect [15, 132]. The essence of iterative learning control is to repeatedly try the same control, with the error to constantly modify the control parameters, through the continuous reduction of error to get the desired control effect. In the finite interval, the iterative learning control can achieve perfect tracking of the output of the nonlinear controlled object [133]. Compared with other control methods, the advantages of iterative learning control is obvious, its learning speed, robustness, convergence, algorithm stability and other properties have advantages. However, short development time and the difficulties of integrating with mainstream control methods are its unfavorable factors.

(8) **Neural Network Control**
Neural networks and its application have become a hot topic, and the study of neural network control in flexible beam systems also gets attention [134, 135]. Neural network control can simulate the human online identification of unknown dynamics. Using the neural network, the control problem can be seen as a pattern recognition problem, and the identified pattern is a "change" signal mapped to a "behavior" signal. Comparing with the traditional control strategies, neural network control only requires relatively little system dynamics information, therefore it is more achievable [136, 137]. The most significant feature of neural network control is the ability to learn and it finally stored in the connection network to achieve through the continuous correction of the connection between the neuron's weights, and discretizing. It has a good influence on the control of nonlinear systems and difficult-to-model systems, therefore neural networks are widely used in the control design of uncertain nonlinear systems [138, 139].

(9) **Fuzzy Logic Control**
Fuzzy logic control is another effective method for systems whose models are difficult to establish [140, 141]. It provides a control law based on the rules of natural language. Because the fuzzy model can approximate the nonlinear

system with arbitrary precision, it is very successful to describe the actual system with strong nonlinearity and uncertainty. In recent years, researchers pay more attention to fuzzy modeling and its applications [142].

1.5 Outline of the Book

The general objectives of the book are to develop constructive and systematic methods of designing control for flexible beam systems. The book starts with a brief introduction of modeling methods and control techniques for classes of flexible beam systems in this chapter.

Chapter 2 presents the preliminaries and several lemmas for the subsequent development to simplify the dynamical modeling and further stability analysis for the beam structures, and the dynamic model is also introduced.

In Chap. 3, we investigate the boundary control design problem for an Euler-Bernoulli beam model under both the unknown spatiotemporally varying distributed disturbance and time-varying boundary disturbance. The adaptive control laws and disturbance observers are proposed to deal with system parametric uncertainties and external disturbances, respectively.

In Chap. 4, the control problem of an Euler-Bernoulli beam with boundary output constraint is addressed. We design a boundary barrier control scheme and apply the proposed Lyapunov function to the original partial differential equations in order to avoid the spillover problem. The form of Lyapunov function combining both the integral Lyapunov function and the barrier Lyapunov function is employed for the control design and stability analysis of the system.

In Chap. 5, the largest challenge is how to eliminate the nonlinear input saturation characteristic and to design an effective active control law for the flexible beam system. We propose the hyperbolic tangent function in the control design of the flexible system to reduce the influence induced by the input saturation.

In Chap. 6, the boundary control architecture is proposed for continuous-time PDE systems preceded by input dead-zone nonlinearity. By using a new description of the dead-zone, boundary control schemes are developed to regulate the deformation of Euler-Bernoulli beam system even in the presence of the external disturbance. It is proved that the proposed control law can ensure uniformly ultimately boundness of the entire system and achieve stabilization.

In Chap. 7, we discuss a problem of a flexible beam system with input backlash characteristic. By transforming the nonlinear input backlash to linear input, the boundary control law and disturbance observer are designed at the top of flexible beam via constructing a Lyapunov candidate function to reduce the vibration of the system. The transverse displacement of the closed-loop system is proved to converge to a small neighborhood of zero.

In Chap. 8, we present the control design for the flexible Euler-Bernoulli beam system with global constraint and uncertain tip payload. To prevent the constraint violation, the Integral- Barrier Lyapunov Function is employed for the control design and stability analysis. We propose a distributed control so that the deflection of the mechanical system can track the desired signal. Exponential stability is well achieved without violation of the constraint.

In Chap. 9, we propose an adaptive boundary iterative learning control (ABILC) scheme for an Euler-Bernoulli beam system with an aperiodic distributed disturbance, an aperiodic boundary disturbance and an unknown system parameter. In order to tackle the input saturation, a hyperbolic tangent function and a saturation function are utilized. Three adaptive laws are proposed and learned along the iteration axis. Based on a time-weighted Lyapunov-Krasovskii-like composite energy function, the restrained ABILC law is designed. For the closed-loop system, the boundedness of all the signals in each iteration is guaranteed. Along the iteration axis, the displacements converge to zero in the presence of the external disturbances.

In Chap. 10, in order to analyze the flexible beam structure, the assumed mode method (AMM) is employed to develop the dynamic model. Based on the N-dimensional discrete dynamic model, neural network control is investigated to track the desired trajectory accurately and to suppress the flexible vibration maximally. To ensure stability rigorously as the goal, the system is proved to be uniform ultimate boundedness (UUB) by Lyapunov stability method. Eventually, simulations verify that the proposed control strategy is effective.

In Chap. 11, a model of a coupled three-dimensional flexible beam with a tip payload is derived by using Hamilton principle and described by a set of partial differential equations and ordinary differential equations. Boundary control laws are proposed to suppress the vibration of the beam in both longitudinal and transverse directions with the environmental disturbance, and the uniform boundedness and the uniformly ultimate boundedness of the closed-loop system are proved. Adaptive control laws are proposed to overcome the uncertainty of the tip payload, and the uniform boundedness and the uniformly ultimate boundedness of the system are proved.

In Chap. 12, the conclusions are summarized and the future work are given.

References

1. W. He, C. Yang, J. Zhu, J.-K. Liu, X. He, Active vibration control of a nonlinear three-dimensional Euler-Bernoulli beam. J. Vib. Control **23**(19), 3196–3215 (2017)
2. W. He, T. Meng, D. Huang, X. Li, Adaptive boundary iterative learning control for an Euler-Bernoulli beam system with input constraint. IEEE Trans. Neural Netw. Learn. Syst. **29**(5), 1539–1549 (2018)
3. B.-Z. Guo, W. Guo, The strong stabilization of a one-dimensional wave equation by non-collocated dynamic boundary feedback control. Automatica **45**(3), 790–797 (2009)
4. B. Bhikkaji, S.O.R. Moheimani, I. Petersen, A negative imaginary approach to modeling and control of a collocated structure. IEEE/ASME Trans. Mechatron. **17**(4), 717–727 (2012)
5. Q.-L. Hu, Z. Wang, H. Gao, Sliding mode and shaped input vibration control of flexible systems. IEEE Trans. Aerosp. Electron. Syst. **44**(2), 503–519 (2008)

6. B.-Z. Guo, F.-F. Jin, The active disturbance rejection and sliding mode control approach to the stabilization of the Euler-Bernoulli beam equation with boundary input disturbance. Automatica **49**(9), 2911–2918 (2013)
7. D. Halim, S. Moheimani, Spatial resonant control of flexible structures-application to a piezoelectric laminate beam. IEEE Trans. Control Syst. Technol. **9**(1), 37–53 (2001)
8. W. He, S.S. Ge, B.V.E. How, Y.S. Choo, K.-S. Hong, Robust adaptive boundary control of a flexible marine riser with vessel dynamics. Automatica **47**(4), 722–732 (2011)
9. A. El Jai, A. Pritchard, Sensors and actuators in distributed systems. Int. J. Control **46**(4), 1139–1153 (1987)
10. Q.C. Nguyen, K.-S. Hong, Asymptotic stabilization of a nonlinear axially moving string by adaptive boundary control. J. Sound Vib. **329**(22), 4588–4603 (2010)
11. W. He, S.S. Ge, Robust adaptive boundary control of a vibrating string under unknown time-varying disturbance. IEEE Trans. Control Syst. Technol. **20**(1), 48–58 (2012)
12. K.-J. Yang, K.-S. Hong, F. Matsuno, Robust boundary control of an axially moving string by using a PR transfer function. IEEE Trans. Autom. Control **50**(12), 2053–2058 (2005)
13. W. He, S. Zhang, S.S. Ge, Boundary output-feedback stabilization of a Timoshenko beam using disturbance observer. IEEE Trans. Ind. Electron. **60**(11), 5186–5194 (2013)
14. Y. Li, D. Aron, C.D. Rahn, Adaptive vibration isolation for axially moving strings: theory and experiment. Automatica **38**(3), 379–390 (2002)
15. D. Huang, J.-X. Xu, Steady-state iterative learning control for a class of nonlinear PDE processes. J. Process Control **21**(8), 1155–1163 (2011)
16. D. Huang, J.-X. Xu, X. Li, C. Xu, M. Yu, D-type anticipatory iterative learning control for a class of inhomogeneous heat equations. Automatica **49**(8), 2397–2408 (2013)
17. C. Xu, E. Schuster, R. Vazquez, M. Krstic, Stabilization of linearized 2d magnetohydrodynamic channel flow by backstepping boundary control. Syst. Control Lett. **57**(10), 805–812 (2008)
18. M. Krstic, A. Smyshlyaev, *Boundary Control of PDEs: A Course on Backstepping Designs* (Society for Industrial and Applied Mathematics, Philadelphia, 2008)
19. A. Smyshlyaev, M. Krstic, *Adaptive Control of Parabolic PDEs* (Princeton University Press, New Jersey, 2010)
20. J.J.E. Slotine, W. Li, *Applied Nonlinear Control* (Prentice Hall Englewood Cliffs, NJ, 1991)
21. F.-F. Jin, B.-Z. Guo, Lyapunov approach to output feedback stabilization for the Euler-Bernoulli beam equation with boundary input disturbance. Automatica **52**(1), 95–102 (2015)
22. T.M. Guerra, L. Vermeiren, Lmi-based relaxed nonquadratic stabilization conditions for nonlinear systems in the takagi-sugeno's form. Automatica **40**(5), 823–829 (2004)
23. H. Lam, H. Li, Output-feedback tracking control for polynomial fuzzy-model-based control systems. IEEE Trans. Ind. Electron. **60**(12), 5830–5840 (2011)
24. S. Labiod, M.S. Boucherit, T.M. Guerra, Adaptive fuzzy control of a class of mimo nonlinear systems. Fuzzy Sets Syst. **151**(1), 59–77 (2005)
25. K.D. Do, J. Pan, Boundary control of transverse motion of marine risers with actuator dynamics. J. Sound Vib. **318**(4), 768–791 (2008)
26. S.K. Chakrabarti, R.E. Frampton, Review of riser analysis techniques. Appl. Ocean Res. **4**(2), 73–90 (1982)
27. C.T. Kiang, A. Spowage, C.K. Yoong, Review of control and sensor system of flexible manipulator. J. Intell. Robot. Syst. **77**(1), 187–213 (2015)
28. J. Liu, W. He, *Distributed Parameter Modeling and Boundary Control of Flexible Manipulators* (Springer, Berlin, 2018)
29. Y. Liu, Z. Zhao, W. He, Boundary control of an axially moving accelerated/decelerated belt system. Int. J. Robust Nonlinear Control **26**(17), 3849–3866 (2016)
30. A.D. Wright, *Modern Control Design for Flexible Wind Turbines* (National Renewable Energy Laboratory, 2004)
31. C. Van der Woude, A study on vibration isolation in a wind turbine subjected to wind and seismic loading. Master's thesis, University of Waterloo, 2011

32. W. He, S. Nie, T. Meng, Y.-J. Liu, Modeling and vibration control for a moving beam with application in a drilling riser. IEEE Trans. Control Syst. Technol. **25**(3), 1036–1043 (2017)
33. K.D. Do, Stochastic boundary control design for extensible marine risers in three dimensional space. Automatica **77**(3), 184–197 (2017)
34. E. Pereira, S.S. Aphale, V. Feliu, S.R. Moheimani, Integral resonant control for vibration damping and precise tip-positioning of a single-link flexible manipulator. IEEE/ASME Trans. Mechatron. **16**(2), 232–240 (2011)
35. X. Zhang, W. Xu, S.S. Nair, V. Chellaboina, PDE modeling and control of a flexible two-link manipulator. IEEE Trans. Control Syst. Technol. **13**(2), 301–312 (2005)
36. S.K. Pradhan, B. Subudhi, Nonlinear adaptive model predictive controller for a flexible manipulator: an experimental study. IEEE Trans. Control Syst. Technol. **22**(5), 1754–1768 (2014)
37. Z. Zhao, Y. Liu, W. He, L. Fei, Adaptive boundary control of an axially moving belt system with high acceleration/deceleration. IET Control Theory Appl. **10**(11), 1299–1306 (2016)
38. Y. Li, C.D. Rahn, Adaptive vibration isolation for axially moving beams. IEEE/ASME Trans. Mechatron. **5**(4), 419–428 (2000)
39. K.-S. Hong, C.-W. Kim, K.-T. Hong, Boundary control of an axially moving belt system in a thin-metal production line. Int. J. Control Autom. Syst. **2**(1), 55–67 (2004)
40. J. Young, S.M. Walker, R.J. Bomphrey, G.K. Taylor, A.L. Thomas, Details of insect wing design and deformation enhance aerodynamic function and flight efficiency. Science **325**(5947), 1549–1552 (2009)
41. W. He, S. Zhang, Control design for nonlinear flexible wings of a robotic aircraft. IEEE Trans. Control Syst. Technol. **25**(1), 351–357 (2017)
42. A.A. Paranjape, J. Guan, S.-J. Chung, M. Krstic, PDE boundary control for flexible articulated wings on a robotic aircraft. IEEE Trans. Robot. **29**(3), 625–640 (2013)
43. A. Chakravarthy, K.A. Evans, J. Evers, Sensitivities and functional gains for a flexible aircraft-inspired model, in *American Control Conference (ACC), 2010* (IEEE, 2010), pp. 4893–4898
44. N. Zimmert, A. Pertsch, O. Sawodny, 2-dof control of a fire-rescue turntable ladder. IEEE Trans. Control Syst. Technol. **20**(2), 438–452 (2012)
45. Q. Yuan, J. Lew, D. Piyabongkarn, Motion control of an aerial work platform, in *American Control Conference (ACC), 2009* (IEEE, 2009), pp. 2873–2878
46. A. Pertsch, N. Zimmert, O. Sawodny, Modeling a fire-rescue turntable ladder as piecewise Euler-Bernoulli beam with a tip mass, in *Proceedings of the 48th IEEE Conference on Decision and Control, 2009 held jointly with the 2009 28th Chinese Control Conference. CDC/CCC 2009* (IEEE, 2009), pp. 7321–7326
47. W. He, S.S. Ge, Vibration control of a nonuniform wind turbine tower via disturbance observer. IEEE/ASME Trans. Mechatron. **20**(1), 237–244 (2015)
48. J. Genov, B. Gilev, Y. Slavchev, G. Venkov, Modeling and control of wind turbine tower vibrations, in *AIP Conference Proceedings*, vol. 1293 (AIP, 2010), pp. 30–38
49. Y. Si, H.R. Karimi, H. Gao, Modeling and parameter analysis of the OC3-hywind floating wind turbine with a tuned mass damper in nacelle. J. Appl. Math. **2013** (2013)
50. S.S. Ge, S. Zhang, W. He, Vibration control of an Euler-Bernoulli beam under unknown spatiotemporally varying disturbance. Int. J. Control **84**(5), 947–960 (2011)
51. N. Wang, H.-N. Wu, L. Guo, Coupling-observer-based nonlinear control for flexible air-breathing hypersonic vehicles. Nonlinear Dyn. **1**(1), 1–24 (2014)
52. J.-W. Wang, H.-X. Li, H.-N. Wu, Distributed proportional plus second-order spatial derivative control for distributed parameter systems subject to spatiotemporal uncertainties. Nonlinear Dyn. **76**(4), 2041–2058 (2014)
53. H.-N. Wu, J.-W. Wang, Observer design and output feedback stabilization for nonlinear multivariable systems with diffusion PDE-governed sensor dynamics. Nonlinear Dyn. **72**(3), 615–628 (2013)
54. W. He, X. He, S.S. Ge, Boundary output feedback control of a flexible string system with input saturation. Nonlinear Dyn. **80**(2), 871–888 (2015)
55. X. Cai, M. Krstic, Nonlinear stabilization through wave PDE dynamics with a moving uncontrolled boundary. Automatica **68**, 27–38 (2016)

56. I. Karafyllis, M. Krstic, Sampled-data boundary feedback control of 1-D parabolic PDEs. Automatica **87**, 226–237 (2018)
57. H. Goldstein, *Classical Mechanics* (Pearson Education India, 1965)
58. O.A. Bauchau, *Flexible Multibody Dynamics*, vol. 176 (Springer Science and Business Media, Berlin, 2010)
59. B. Kang, J.K. Mills, Dynamic modeling of structurally-flexible planar parallel manipulator. Robotica **20**(03), 329–339 (2002)
60. W. Schiehlen, *Multibody Systems Handbook*, vol. 6 (Springer, Berlin, 1990)
61. J. Gibson, A. Adamian, Approximation theory for linear-quadratic-Gaussian control of flexible structures. SIAM J. Control Optim. **29**(1), 1–37 (1991)
62. G.R. Fowles, G.L. Cassiday, *Analytical Mechanics* (Saunders College, 1999)
63. F.P. Beer, E.R. Johnston Jr, D.F. Mazurek, P.J. Cornwell, E.R. Eisenberg, S. Sanghi, *Vector Mechanics for Engineers*, vol. 1 (Tata McGraw-Hill Education, 1972)
64. Z. Liu, J.-K. Liu, W. He, Modeling and vibration control of a flexible aerial refueling hose with variable lengths and input constraint. Automatica **77**(3), 302–310 (2017)
65. W. He, S.S. Ge, Vibration control of a flexible beam with output constraint. IEEE Trans. Ind. Electron. **62**(8), 5023–5030 (2015)
66. X. He, W. He, C. Sun, Robust adaptive vibration control for an uncertain flexible Timoshenko robotic manipulator with input and output constraints. Int. J. Syst. Sci. **48**(13), 2860–2870 (2017)
67. W. He, X. He, S.S. Ge, Vibration control of flexible marine riser systems with input saturation. IEEE/ASME Trans. Mechatron. **21**(1), 254–265 (2016)
68. Y.-L. Hwang, Recursive Newton-Euler formulation for flexible dynamic manufacturing analysis of open-loop robotic systems. Int. J. Adv. Manuf. Technol. **29**(5), 598–604 (2006)
69. P.F. Pai, T.J. Anderson, E.A. Wheater, Large-deformation tests and total-Lagrangian finite-element analyses of flexible beams. Int. J. Solids Struct. **37**(21), 2951–2980 (2000)
70. W.J. Book, Recursive Lagrangian dynamics of flexible manipulator arms. Int. J. Robot. Res. **3**(3), 87–101 (1984)
71. W. Chen, Dynamic modeling of multi-link flexible robotic manipulators. Comput. Struct. **79**(2), 183–195 (2001)
72. X. Zhang, Y. Huang, X. Chen, W. Han, Modeling of a space flexible probe-cone docking system based on the Kane method. Chin. J. Aeronaut. **27**(2), 248–258 (2014)
73. J.W. Gibbs, On the fundamental formulae of dynamics. Am. J. Math. **2**(1), 49–64 (1879)
74. L.A. Pars, *A Treatise on Analytical Dynamics* (Wiley, New York, 1965)
75. Y. Li, B. Tang, Z. Shi, Y. Lu, Experimental study for trajectory tracking of a two-link flexible manipulator. Int. J. Syst. Sci. **31**(1), 3–9 (2000)
76. T. Zebin, M. Alam, Modeling and control of a two-link flexible manipulator using fuzzy logic and genetic optimization techniques. J. Comput. **7**(3), 578–585 (2012)
77. Z. Razali, D. Derin, N. Atikah, Finite element analysis on robotic arm for waste management application. Appl. Mech. Mater. **786**, 372–377 (2015)
78. M.M. Kokurin, Necessary and sufficient conditions for the polynomial convergence of the quasi-reversibility and finite-difference methods for an ill-posed cauchy problem with exact data. Comput. Math. Math. Phys. **55**(12), 1986–2000 (2015)
79. H. Mirahki, M. Moallem, S.A. Rahimi, Design optimization of ipmsm for 42 V integrated starter alternator using lumped parameter model and genetic algorithms. IEEE Trans. Magn. **50**(3), 114–119 (2014)
80. A. Green, J. Sasiadek, Robot manipulator control for rigid and assumed mode flexible dynamics models, in *AIAA Guidance, Navigation, and Control Conference and Exhibit* (2003), p. 5435
81. W. He, Y. Ouyang, J. Hong, Vibration control of a flexible robotic manipulator in the presence of input deadzone. IEEE Trans. Ind. Inform. **13**(1), 48–59 (2017)
82. D.L. Logan, *A First Course in the Finite Element Method* (Cengage Learning, Boston, 2011)
83. G.W. Recktenwald, Finite-difference approximations to the heat equation. Mech. Eng. **10**, 1–27 (2004)

84. C. Monje, F. Ramos, V. Feliu, B. Vinagre, Tip position control of a lightweight flexible manipulator using a fractional order controller. IET Control Theory Appl. **1**(5), 1451–1460 (2007)
85. A. Shawky, A. Ordys, L. Petropoulakis, M. Grimble, Position control of flexible manipulator using non-linear H with state-dependent Riccati equation. Proc. Inst. Mech. Eng. Part I: J. Syst. Control Eng. **221**(3), 475–486 (2007)
86. J.F.P. Solís, G.S. Navarro, R.C. Linares, Modeling and tip position control of a flexible link robot: experimental results. Computación y Sistemas **12**(4), 421–435 (2009)
87. R. Morales, V. Feliu, V. Jaramillo, Position control of very lightweight single-link flexible arms with large payload variations by using disturbance observers. Robot. Auton. Syst. **60**(4), 532–547 (2012)
88. A. Bazaei, M. Moallem, Improving force control bandwidth of flexible-link arms through output redefinition. IEEE/ASME Trans. Mechatron. **16**(2), 380–386 (2011)
89. I. Payo, V. Feliu, O.D. Cortázar, Force control of a very lightweight single-link flexible arm based on coupling torque feedback. Mechatronics **19**(3), 334–347 (2009)
90. A. Garcia, V. Feliu, Force control of a single-link flexible robot based on a collision detection mechanism. IEE Proc. Control Theory Appl. **147**(6), 588–595 (2000)
91. Z.-C. Qiu, X.-M. Zhang, Y.-C. Wang, Z.-W. Wu, Active vibration control of a flexible beam using a non-collocated acceleration sensor and piezoelectric patch actuator. J. Sound Vib. **326**(3), 438–455 (2009)
92. A. Jnifene, Active vibration control of flexible structures using delayed position feedback. Syst. Control Lett. **56**(3), 215–222 (2007)
93. Z. Mohamed, J. Martins, M. Tokhi, J.S. Da Costa, M. Botto, Vibration control of a very flexible manipulator system. Control Eng. Pract. **13**(3), 267–277 (2005)
94. M.M. Zain, M. Tokhi, Z. Mohamed, Hybrid learning control schemes with input shaping of a flexible manipulator system. Mechatronics **16**(3), 209–219 (2006)
95. M.S. Alam, M.O. Tokhi, Hybrid fuzzy logic control with genetic optimisation for a single-link flexible manipulator. Eng. Appl. Artif. Intell. **21**(6), 858–873 (2008)
96. M.A. Ahmad, A.N.K. Nasir, R.R. Ismail, M.S. Ramli, Comparison of hybrid control schemes for vibration suppression of flexible robot manipulator, in *2009 International Conference on Computer Modeling and Simulation (ICCMS)* (IEEE, 2009), pp. 356–360
97. T. Sales, D. Rade, L. De Souza, Passive vibration control of flexible spacecraft using shunted piezoelectric transducers. Aerosp. Sci. Technol. **29**(1), 403–412 (2013)
98. W.D. Zhu, J. Ni, J. Huang, Active control of translating media with arbitrarily varying length. J. Vib. Acoust. **123**(3), 347–358 (2001)
99. H. Feng, B.-Z. Guo, Output feedback stabilization of an unstable wave equation with general corrupted boundary observation. Automatica **50**(12), 3164–3172 (2014)
100. W. He, S. Zhang, S.S. Ge, Robust adaptive control of a thruster assisted position mooring system. Automatica **50**(7), 1843–1851 (2014)
101. T. Nguyen, K.D. Do, J. Pan, Boundary control of two-dimensional marine risers with bending couplings. J. Sound Vib. **332**(16), 3605–3622 (2013)
102. A.A. Siranosian, M. Krstic, A. Smyshlyaev, M. Bement, Motion planning and tracking for tip displacement and deflection angle for flexible beams. J. Dyn. Syst. Meas. Control **131**(3), 031009 (2009)
103. M. Krstic, *Boundary Control of PDEs: A Course on Backstepping Designs* (Society for Industrial and Applied Mathematics, Philadelphia, 2008)
104. O. Morgul, Dynamic boundary control of a Euler-Bernoulli beam. IEEE Trans. Autom. Control **37**(5), 639–642 (1992)
105. X. He, W. He, J. Shi, C. Sun, Boundary vibration control of variable length crane systems in two-dimensional space with output constraints. IEEE/ASME Trans. Mechatron. **22**(5), 1952–1962 (2017)
106. K. Endo, F. Matsuno, H. Kawasaki, Simple boundary cooperative control of two one-link flexible arms for grasping. IEEE Trans. Autom. Control **54**(10), 2470–2476 (2009)

107. G.Q. Xu, H. Wang, Stabilisation of Timoshenko beam system with delay in the boundary control. Int. J. Control **86**(6), 1165–1178 (2013)
108. F. Conrad, Stabilization of beams by pointwise feedback control. Siam J. Control Optim. **28**(2), 423–437 (2006)
109. K. Ammari, M. Tucsnak, Stabilization of Bernoulli-Euler beams by means of a pointwise feedback force. SIAM J. Control Optim. **39**(4), 1160–1181 (2000)
110. G.Q. Xu, S.P. Yung, Stabilization of Timoshenko beam by means of pointwise controls. Esaim Control Optim. Calc. Var. **9**(9), 579–600 (2003)
111. B. Bamieh, F. Paganini, M.A. Dahleh, Distributed control of spatially invariant systems. IEEE Trans. Autom. Control **47**(7), 1091–1107 (2002)
112. F. Wu, Distributed control for interconnected linear parameter-dependent systems. IEE Proc. Control Theory Appl. **150**(5), 518–527 (2003)
113. O.J. Aldraihem, R.C. Wetherhold, T. Singh, Distributed control of laminated beams: Timoshenko theory versus Euler-Bernoulli theory. J. Intell. Mater. Syst. Struct. **8**(2), 149–157 (1997)
114. W. He, C. Yang, T. Meng, C. Sun, Distributed control of a class of flexible mechanical systems with global constraint. Int. J. Control **89**(1), 128–139 (2016)
115. G.K. Mann, B.-G. Hu, R.G. Gosine, Analysis of direct action fuzzy PID controller structures. IEEE Trans. Syst. Man Cybern. Part B (Cybern.) **29**(3), 371–388 (1999)
116. M.M. Jovanović, A.M. Simonović, N.D. Zorić, N.S. Lukić, S.N. Stupar, S.S. Ilić, Experimental studies on active vibration control of a smart composite beam using a PID controller. Smart Mater. Struct. **22**(11), 115038 (2013)
117. H.-Y. Zhu, H.-N. Wu, J.-W. Wang, H_∞ disturbance attenuation for nonlinear coupled parabolic PDE-ODE systems via fuzzy-model-based control approach. IEEE Trans. Syst. Man Cybern. Syst. **47**(8), 1814–1825 (2017)
118. M. KrstiC, J. Sun, P.V. Kokotovic, Robust control of nonlinear systems with input unmodeled dynamics. IEEE Trans. Autom. Control **41**(6), 913–920 (1996)
119. S.S. Ge, T.H. Lee, G. Zhu, Energy-based robust controller design for multi-link flexible robots. Mechatronics **6**(7), 779–798 (1996)
120. S.S. Ge, T.H. Lee, Z. Wang, Model-free regulation of multi-link smart materials robots. IEEE/ASME Trans. Mechatron. **6**(3), 346–351 (2001)
121. R.J. Theodore, A. Ghosal, Robust control of multilink flexible manipulators. Mech. Mach. Theory **38**(4), 367–377 (2003)
122. L. Tian, C. Collins, Adaptive neuro-fuzzy control of a flexible manipulator. Mechatronics **15**(10), 1305–1320 (2005)
123. W. Guo, B.-Z. Guo, Parameter estimation and non-collocated adaptive stabilization for a wave equation subject to general boundary harmonic disturbance. IEEE Trans. Autom. Control **58**(7), 1631–1643 (2013)
124. K.-J. Yang, K.-S. Hong, F. Matsuno, Energy-based control of axially translating beams: varying tension, varying speed, and disturbance adaptation. IEEE Trans. Control Syst. Technol. **13**(6), 1045–1054 (2005)
125. C.L.P. Chen, G.-X. Wen, Y.-J. Liu, Z. Liu, Observer-based adaptive backstepping consensus tracking control for high-order nonlinear semi-strict-feedback multiagent systems. IEEE Trans. Cybern. **46**(7), 1591–1601 (2016)
126. T. Lee, S. Ge, Z. Wang, Adaptive robust controller design for multi-link flexible robots. Mechatronics **11**(8), 951–967 (2001)
127. K.B. Ariyur, M. Krstic, *Real-Time Optimization by Extremum-Seeking Control* (Wiley, New York, 2003)
128. F.-Y. Wang, J. Zhang, Q. Wei, X. Zheng, L. Li, PDP: parallel dynamic programming. IEEE/CAA J. Autom. Sin. **4**(1), 1–5 (2017)
129. G. Zhu, S.S. Ge, A quasi-tracking approach for finite-time control of a mass-beam system. Automatica **34**(7), 881–888 (1998)
130. S.S. Ge, T.H. Lee, G. Zhu, F. Hong, Variable structure control of a distributed-parameter flexible beam. J. Robot. Syst. **18**(1), 17–27 (2001)

131. L. Chen, Y. Yan, C. Mu, C. Sun, Characteristic model-based discrete-time sliding mode control for spacecraft with variable tilt of flexible structures. IEEE/CAA J. Autom. Sin. **3**(1), 42–50 (2016)
132. M.A. Fadil, N.A. Jalil, I.Z.M. Darus, Intelligent PID controller using iterative learning algorithm for active vibration controller of flexible beam, in *2013 IEEE Symposium on Computers and Informatics (ISCI)* (IEEE, 2013), pp. 80–85
133. D. Huang, X. Li, J.-X. Xu, C. Xu, W. He, Iterative learning control of inhomogeneous distributed parameter systems frequency domain design and analysis. Syst. Control Lett. **72**, 22–29 (2014)
134. X. Cheng, R. Patel, Neural network based tracking control of a flexible macro-micro manipulator system. Neural Netw. **16**(2), 271–286 (2003)
135. S.S. Ge, T.H. Lee, G. Zhu, A nonlinear feedback controller for a single-link flexible manipulator based on a finite element model. J. Robot. Syst. **14**(3), 165–178 (1997)
136. M. Moradi, H. Malekizade, Neural network identification based multivariable feedback linearization robust control for a two-link manipulator. J. Intell. Robot. Syst. **72**(2), 167–178 (2013)
137. H. Yang, J. Liu, An adaptive RBF neural network control method for a class of nonlinear systems. IEEE/CAA J. Autom. Sin. **5**(2), 457–462 (2018)
138. C.L.P. Chen, G.-X. Wen, Y.-J. Liu, F.-Y. Wang, Adaptive consensus control for a class of nonlinear multiagent time-delay systems using neural networks. IEEE Trans. Neural Netw. Learn. Syst. **25**(6), 1217–1226 (2014)
139. H.T. Dinh, R. Kamalapurkar, S. Bhasin, W.E. Dixon, Dynamic neural network-based robust observers for uncertain nonlinear systems. Neural Netw. **60**, 44–52 (2014)
140. A. Jnifene, W. Andrews, Experimental study on active vibration control of a single-link flexible manipulator using tools of fuzzy logic and neural networks. IEEE Trans. Instrum. Meas. **54**(3), 1200–1208 (2005)
141. A. Green, J. Sasiadek, Adaptive control of a flexible robot using fuzzy logic. J. Guid. Control Dyn. **28**(1), 36–42 (2005)
142. A. Chatterjee, R. Chatterjee, F. Matsuno, T. Endo, Augmented stable fuzzy control for flexible robotic arm using LMI approach and neuro-fuzzy state space modeling. IEEE Trans. Ind. Electron. **55**(3), 1256–1270 (2008)

Chapter 2
Mathematical Preliminaries

2.1 The Hamilton Principle

As opposed to lumped mechanical systems, flexible beam systems have an infinite number of degrees of freedom, and the model of the system is described by using continuous functions of space and time. The Hamilton principle permits the derivation of equations of motion from energy quantities in a variational form and generates the motion equations of the flexible beam systems. The Hamilton principle [1, 2] is represented by

$$\int_{t_1}^{t_2} \delta(E_k - E_p + W)dt = 0 \tag{2.1}$$

where t_1 and t_2 are two time instants, $t_1 < t < t_2$ is the operating interval, δ denotes the variational operator, E_k and E_p are the kinetic and potential energies of the system, respectively, and W denotes the work done by the nonconservative forces acting on the system, including internal tension, transverse load, linear structural damping, and external disturbance. The principle states that the variation of the kinetic and potential energies plus the variation of work done by loads during any time interval $[t_1, t_2]$ must be equal to zero.

There are some advantages using the Hamilton principle to derive the mathematical model of the flexible beam systems. Firstly, this approach is independent of the coordinates, and the boundary conditions can be automatically generated by this approach. In addition, the kinetic energy, the potential energy, and the work done by the nonconservative forces in the Hamilton principle can be directly used to design the Lyapunov function of the closed-loop system.

W. He and J. Liu, *Active Vibration Control and Stability Analysis of Flexible Beam Systems*, https://doi.org/10.1007/978-981-10-7539-1_2

2.2 Lyapunov Stability Theorem

Lyapunov design has been a primary tool for nonlinear control system design, stability and performance analysis since its introduction in 1982. The basic idea is to design a feedback control law that renders the derivative of a specified Lyapunov function candidate negative definite or negative semi-definite. Lyapunov's direct method is a mathematical interpretation of the physical property that if a system's total energy is dissipating, then the states of the system will ultimately reach an equilibrium point. The basic idea behind the method is that, if there exist a kind of continuous scalar 'energy' functions such that this 'energy' diminishes along the system's trajectory, then the system is said to be asymptotically stable. Since there is no need to solve the solution of the differential equations governing the system in determining its stability, it is usually referred to as the direct method [3].

Although Lyapunov's direct method is efficient for stability analysis, its applicability is restricted due to the difficulty in selecting a Lyapunov function. The situation is different when facing the controller design problem, where the control has not been specified, and the system under consideration is undetermined. Lyapunov functions have been effectively utilized in the synthesis of control systems. The basic idea is that, by first choosing a Lyapunov function candidate and then the feedback control law can be specified such that it renders the derivative of the specified Lyapunov function candidate negative definite, or negative semi-definite when invariance principle can be used to prove asymptotic stability. This way of designing control is called Lyapunov design. Lyapunov design depends on the selection of Lyapunov function candidates. Though the result is sufficient, it is difficult to find a Lyapunov function (LF) satisfying the requirements of Lyapunov design. Fortunately, during the past several decades, many effective control design approaches have been developed for different classes of linear and nonlinear systems based on the basic ideas of Lyapunov design. Lyapunov functions are additive, like energy, i.e., Lyapunov functions for combinations of subsystems may be derived by adding the Lyapunov functions of the subsystems.

Though Lyapunov design is a very powerful tool for control system design, stability and performance analysis, the construction of a Lyapunov function is not easy for general nonlinear systems, and it is usually a trial-and-error process and there is a lack of systematic methods. Different choices of Lyapunov functions may result in different control structures and control performance. Past experience shows that a good design of Lyapunov function should fully utilize the property of the studied systems.

In actual applications, Lyapunov design can be conceptually divided into two steps,

(1) choose a candidate Lyapunov function V for the system;
(2) design a controller which renders its derivative $\dot{V}$ negative.

Sometimes, it may be more advantageous to reverse the order of operation, i.e., designing a controller that is most likely to be able to stabilize the closed-loop system first by examining the properties of the system, and then choose a Lyapunov function candidate V for the closed-loop system to show that it is indeed a Lyapunov function. Lyapunov design is sufficient. Stabilizing controllers are obtained if the processes succeed. If the attempts fail, no conclusion can be drawn on the existence of a stabilizing controller.

Let function $V(x)$ be a Lyapunov function candidate [4]. Thus the task is to search for the function to guarantee that, for all $x \in R^n$, the time derivative of $V(x)$ satisfies

$$\dot{V}(x) \leq -W(x) \tag{2.2}$$

where $W(x)$ is a positive definite function.

In general, this is a difficult task. A system for which a good choice of $V(x)$ and $W(x)$ exists is said to possess a control Lyapunov function. As a design tool for general nonlinear systems, the main deficiency is that the Lyapunov function is unknown. The task of finding an appropriate Lyapunov function may be as complex as that of designing a stabilizing feedback law. However, for flexible beam systems, these two tasks can be solved simultaneously. It should be noted that there may exist many Lyapunov functions for a given nonlinear system. Specific choices of Lyapunov functions may yield better and cleaner controllers than others.

2.3 Useful Lemmas

Remark 2.1 Throughout the book, we give the definitions such that $(*)' = \frac{\partial(*)}{\partial x}$, $(*)'' = \frac{\partial^2(*)}{\partial x^2}$, $(*)''' = \frac{\partial^3(*)}{\partial x^3}$, $(*)'''' = \frac{\partial^4(*)}{\partial x^4}$, $\dot{(*)} = \frac{\partial(*)}{\partial t}$ and $\ddot{(*)} = \frac{\partial^2(*)}{\partial t^2}$.

We present the following lemma for the subsequent development.

Lemma 2.1 ([5]) *Let $\phi_1(x,t)$, $\phi_2(x,t) \in \mathbb{R}$ with $x \in [0, L]$ and $t \in [0, \infty)$, the following inequalities hold*

$$\phi_1(x,t)\phi_2(x,t) \leq |\phi_1(x,t)\phi_2(x,t)| \leq \phi_1^2(x,t) + \phi_2^2(x,t), \quad \forall \phi_1(x,t), \phi_2(x,t) \in \mathbb{R}. \tag{2.3}$$

Lemma 2.2 ([5]) *Let $\phi_1(x,t)$, $\phi_2(x,t) \in \mathbb{R}$ with $x \in [0, L]$ and $t \in [0, \infty)$, the following inequalities hold*

$$|\phi_1(x,t)\phi_2(x,t)| = \left|\left(\frac{1}{\sqrt{\delta}}\phi_1(x,t)\right)(\sqrt{\delta}\phi_2(x,t))\right| \leq \frac{1}{\delta}\phi_1^2(x,t) + \delta\phi_2^2(x,t), \tag{2.4}$$

$\forall \phi_1(x,t), \phi_2(x,t) \in \mathbb{R}$ *and* $\delta > 0$.

Lemma 2.3 ([6]) *Rayleigh–Ritz theorem: Let $A \in \mathbb{R}^{n\times n}$ be a real, symmetric, positive-definite matrix, therefore, all the eigenvalues of A are real and positive. Let $\lambda_{\min}$ and $\lambda_{\max}$ denote the minimum and maximum eigenvalues of A, respectively, then for $\forall x \in \mathbb{R}^n$, we have*

$$\lambda_{\min}||x||^2 \leq x^T Ax \leq \lambda_{\max}||x||^2, \tag{2.5}$$

where $||\cdot||$ denotes the standard Euclidean norm.

Lemma 2.4 ([7]) *Let $\phi(x,t) \in \mathbb{R}$ be a function defined on $x \in [0, L]$ and $t \in [0, \infty)$ that satisfies the boundary condition*

$$\phi(0,t) = 0, \quad \forall t \in [0, \infty), \tag{2.6}$$

then the following inequality holds

$$\phi^2(x,t) \leq L \int_0^L [\phi'(x,t)]^2 dx, \quad \forall x \in [0, L]. \tag{2.7}$$

If in addition to (2.6), the function $\phi(x,t)$ satisfies the boundary condition

$$\phi'(0,t) = 0, \quad \forall t \in [0, \infty), \tag{2.8}$$

then the following inequality also holds

$$[\phi'(x,t)]^2 \leq L \int_0^L [\phi''(x,t)]^2 dx, \quad \forall x \in [0, L]. \tag{2.9}$$

2.4 Modeling of a Flexible Beam System

Figure 2.1 shows an Euler–Bernoulli beam system extracted from a class of flexible systems under unknown distributed spatiotemporally varying disturbance $f(x,t)$ and unknown time-varying boundary disturbance $d(t)$. The left boundary of the beam is fixed at origin. Hamilton principle [9] which is used to derive the model of the beam system is represented by

$$\int_{t_1}^{t_2} \delta[E_k(t) - E_p(t) + W(t)]dt = 0, \tag{2.10}$$

where t_1 and t_2 are two time instants, $t_1 < t < t_2$ is the operating interval and δ denotes the variational operator.

The kinetic energy of the beam $E_k(t)$ can be represented as

$$E_k(t) = \frac{1}{2}M\,[\dot{w}(L,t)]^2 + \frac{1}{2}\rho \int_0^L [\dot{w}(x,t)]^2\,dx, \tag{2.11}$$

where x and t represent the independent spatial and time variables respectively.

The potential energy $E_p(t)$ due to the bending can be obtained from

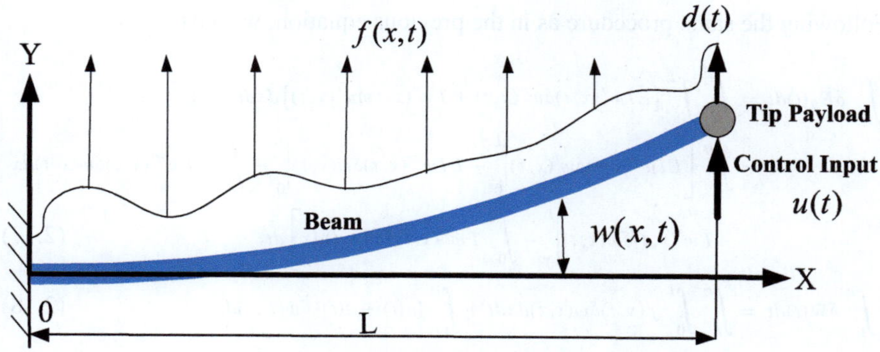

Fig. 2.1 A typical cantilevered beam system

$$E_p(t) = \frac{1}{2}EI\int_0^L \left[w''(x,t)\right]^2 dx + \frac{1}{2}T\int_0^L \left[w'(x,t)\right]^2 dx, \tag{2.12}$$

The virtual work done by disturbances including distributed disturbance $f(x,t)$ on the beam and the boundary disturbance $d(t)$ on the tip payload is given by

$$\delta W_d(t) = \int_0^L f(x,t)\delta w(x,t)dx + d(t)\delta w(L,t). \tag{2.13}$$

The virtual work done by the control input force $u(t)$ which produces a transverse force for vibration suppression can be written as

$$\delta W_f(t) = u(t)\delta w(L,t). \tag{2.14}$$

Then, we have the total virtual work done on the system as

$$\begin{aligned} \delta W(t) &= \delta W_d(t) + \delta W_f(t) \\ &= \int_0^L f(x,t)\delta w(x,t)dx + \left[u(t) + d(t)\right]\delta w(L,t). \end{aligned} \tag{2.15}$$

Applying the variation operator and integrating (2.11), (2.12), and (2.14) by parts respectively, we obtain

$$\begin{aligned} \int_{t_1}^{t_2} \delta E_k(t)dt &= M\int_{t_1}^{t_2} \dot{w}(L,t)\delta\dot{w}(L,t)dt + \rho\int_{t_1}^{t_2}\int_0^L \dot{w}(x,t)\delta\dot{w}(x,t)dxdt \\ &= M\dot{w}(L,t)\delta w(L,t)\Big|_{t_1}^{t_2} - M\int_{t_1}^{t_2} \ddot{w}(L,t)\delta w(L,t)dt \\ &\quad + \rho\int_0^L \left[\dot{w}(x,t)\delta w(x,t)\Big|_{t_1}^{t_2}\right]dx - \rho\int_0^L\int_{t_1}^{t_2} \ddot{w}(x,t)\delta w(x,t)dtdx \\ &= -M\int_{t_1}^{t_2} \ddot{w}(L,t)\delta w(L,t)dt - \rho\int_{t_1}^{t_2}\int_0^L \ddot{w}(x,t)\delta w(x,t)dxdt. \end{aligned} \tag{2.16}$$

Following the same procedure as in the previous equation, we have

$$\begin{aligned}
\int_{t_1}^{t_2} \delta E_p(t)dt &= \int_{t_1}^{t_2} \int_0^L \left[EIw''(x,t)\delta w''(x,t) + Tw'(x,t)\delta w'(x,t)\right] dxdt \\
&= \int_{t_1}^{t_2} \left[EIw''(x,t)\delta w'(x,t)\Big|_0^L - EIw'''(x,t)\delta w(x,t)\Big|_0^L + \int_0^L EIw''''(x,t)\delta w(x,t)dx \right. \\
&\quad \left. + Tw'(x,t)\delta w(x,t)\Big|_0^L - \int_0^L Tw''(x,t)\delta w(x,t)dx \right] dt, \qquad (2.17)
\end{aligned}$$

$$\int_{t_1}^{t_2} \delta W(t)dt = \int_{t_1}^{t_2} \int_0^L f(x,t)\delta w(x,t)dxdt + \int_{t_1}^{t_2} [u(t)+d(t)]\,\delta w(L,t)dt. \qquad (2.18)$$

Substituting (2.16), (2.17) and (2.18) into Hamilton principle (2.10), we obtain

$$\begin{aligned}
&-\int_{t_1}^{t_2} \int_0^L \left[\rho\ddot{w}(x,t) + EIw''''(x,t) - Tw''(x,t) - f(x,t)\right] \delta w(x,t)dxdt \\
&-\int_{t_1}^{t_2} \left[EIw''(x,t)\delta w'(x,t)\Big|_0^L \right] dt - \int_{t_1}^{t_2} \left\{ \left[-EIw'''(x,t) + Tw'(x,t)\right] \delta w(x,t)\Big|_0^L \right. \\
&\left. - [u(t) + d(t) - M\ddot{w}(L,t)]\,\delta w(L,t) \right\} dt = 0. \qquad (2.19)
\end{aligned}$$

As $\delta w(x,t)$ is assumed to be an nonzero arbitrary variation in $0 < x < L$, the expressions under the double integral in (2.19) are set equal to zero. Hence, we obtain the governing equations of the system as

$$\rho\ddot{w}(x,t) + EIw''''(x,t) - Tw''(x,t) = f(x,t), \qquad (2.20)$$

$\forall (x,t) \in (0,L) \times [0,\infty)$. Setting the terms with single integrals in (2.19) equal to zero, we obtain the boundary conditions of the system as

$$\begin{aligned}
w'(0,t) &= 0, \qquad (2.21) \\
w''(L,t) &= 0, \qquad (2.22) \\
w(0,t) &= 0, \qquad (2.23) \\
-EIw'''(L,t) + Tw'(L,t) &= u(t) + d(t) - M\ddot{w}(L,t), \qquad (2.24)
\end{aligned}$$

$\forall t \in [0,\infty)$.

Remark 2.2 With consideration of unknown distributed disturbance $f(x,t)$, the governing equation of the beam system (2.20) is described as a nonhomogeneous PDE. The model of this chapter differs from the beam system governed by a homogeneous PDE in [10], where a proper gain kernel needs to be found to design the backstepping boundary controller and observer. As a consequence, the control schemes in these papers may be not suitable for the system proposed in this chapter.

References

1. H. Goldstein, *Classical Mechanics* (Pearson Education India, 1965)
2. L. Meirovitch, in *Analytical Methods in Vibrations*, vol. 438 (Macmillan, New York, 1967)
3. I.D. Landau, R. Lozano, M. M'Saad, A. Karimi, in *Adaptive Control*, vol. 51 (Springer, Berlin, 1998)
4. H.K. Khalil, in *Noninear Systems*, vol. 2, no. 5 (Prentice-Hall, New Jersey, 1996), pp. 1–5
5. C. Rahn, *Mechatronic Control of Distributed Noise and Vibration* (Springer, New York, 2001)
6. R. Horn, C. Johnson, *Matrix Analysis* (Cambridge University Press, Cambridge, 1990)
7. G.H. Hardy, J.E. Littlewood, G. Polya, *Inequalities* (Cambridge University Press, Cambridge, 1959)
8. E. Hewitt, K. Stromberg, *Real and Abstract Analysis: A Modern Treatment of the Theory of Functions of a Real Variable* (Springer, Berlin, 1929)
9. H. Goldstein, *Classical Mechanics* (Addison-Wesley, Massachusetts, 1951)
10. M. Krstic, A. Smyshlyaev, *Boundary Control of PDEs: A Course on Backstepping Designs* (Society for Industrial and Applied Mathematics, Philadelphia, 2008)

Chapter 3
Vibration Control of a Flexible Beam

3.1 Introduction

An Euler–Bernoulli beam [1–3] is a model that can be used to describe many mechanical flexible systems such as flexible robotic manipulator [4], flexible spacecraft [5], flexible marine riser [6] and moving strip [7]. These physical processes, cannot be modeled by ODEs since the state of the system depends on more than one independent variable [8]. In practice, most of the flexible systems exhibit vibration in the presence of disturbances. Since vibration has been known to reduce the system quality, lead to limited productivity and result in premature fatigue failure, vibration suppression of the beam-like structure has broad applications and has received great attention in the recent years. The chapter is motivated by the industrial applications in boundary control of vibrating flexible structures. Examples of practical applications where vibrating beams are exposed to undesirable spatiotemporally varying disturbances include flexible production risers used for offshore oil transportation, free hanging underwater pipelines, and drilling pipe for drilling mud transportation. The unknown spatiotemporally varying disturbances of the beam lead to the appearance of oscillations, which make the control problems of such systems relatively difficult.

Mathematically, the beam with a tip payload is represented by a set of partial differential equations, i.e., PDEs describing the dynamics of the beam coupled with a set of ordinary differential equations, i.e., ODEs describing the dynamics. The dynamics of the flexible mechanical system modeled by a hybrid model containing both PDEs and ODEs is difficult to control due to the infinite dimensionality of the system. Approaches to control infinite dimensional flexible systems such as the finite element method and the assumed modes method [9–11] are based on the truncated finite-dimensional models of the system. The truncated models are obtained via the model analysis or spatial discretization, in which the flexibility is represented by a finite number of modes by neglecting the higher frequency modes. Spillover problem [12, 13] due to truncation of the model, can lead to an unstable system, which should be avoided.

W. He and J. Liu, *Active Vibration Control and Stability Analysis of Flexible Beam Systems*, https://doi.org/10.1007/978-981-10-7539-1_3

In order to avoid the spillover phenomenon, boundary control [14–20] combining with several other control techniques such as sliding mode control [21], variable structure control [22], energy-based robust control [23], model-free control [24], averaging method [25], has been developed for various infinite dimensional systems. In these approaches, system dynamics analysis and control design are carried out directly based on the original infinite dimensional model. Compared with distributed controllers [26–28], boundary control has several merits. Firstly, boundary control is an economical method to control distributed parameter system without decomposing the system into finite dimensional space. Distributed actuators and sensors are usually needed for the implementation of distributed control. Boundary control is considered to be more practical in vibration control of flexible structures, which requires few actuators and sensors. In addition, boundary control can be derived from a Lyapunov function which is relevant to the mechanical energy based on the dynamics of the system. Due to these advantages, boundary control has gained increasing attention in literatures [6, 14, 16, 17, 29–38]. However, in most of these works, control is designed by neglecting the unknown spatiotemporally varying disturbance.

In this chapter, we investigate the boundary control problem for an Euler–Bernoulli beam model with system parametric uncertainties and under both the unknown spatiotemporally varying distributed disturbance and unknown time-varying boundary disturbance. When bounds of the external disturbances are not available, the control problems become even more difficult.

The rest of the chapter is arranged as follows. The governing equation (PDE) and boundary conditions (ODEs) of an Euler–Bernoulli beam system are introduced by using of Hamilton principle in Sect. 3.2. Boundary control via Lyapunov synthesis is developed in Sect. 3.3, where it is shown that the uniform ultimate boundedness of the closed-loop system can be achieved by the proposed control. Simulations are provided to illustrate the performance of the proposed control in Sect. 3.4. The conclusion of this chapter is presented in Sect. 3.5.

3.2 Problem Formulation

Figure 2.1 shows an Euler–Bernoulli beam model extracted from a class of flexible systems under unknown distributed spatiotemporally varying disturbance $f(x, t)$ and unknown time-varying boundary disturbance $d(t)$. The left boundary of the beam is fixed at origin.

Remark 3.1 With consideration of unknown distributed disturbance $f(x, t)$, the governing equation of the beam system (3.1) is described as a nonhomogeneous PDE. The model of this chapter differs from the beam system governed by a homogeneous PDE in [17], where a proper gain kernel needs to be found to design the backstepping boundary controller and observer. As a consequence, the control schemes in these papers may be not suitable for the system proposed in this chapter. In this chapter, we are going to design the boundary control based on the original nonhomogeneous PDE of the beam system.

Assumption 3.1 For the unknown disturbances $f(x,t)$ and $d(t)$, we assume that there exist constants $\bar{f} \in R^+$ and $\bar{d} \in R^+$, such that $|f(x,t)| \leq \bar{f}, \forall (x,t) \in [0, L] \times [0, \infty)$ and $|d(t)| \leq \bar{d}, \forall (t) \in [0, \infty)$.

Remark 3.2 The unknown time-varying boundary disturbance $d(t)$ which affects the performance of the flexible beam is assumed to be bounded for the finite energy, i.e., $d(t) \in \mathcal{L}_\infty$. In addition, the boundary disturbance is supposed to vary slowly, namely, $\dot{d}(t) = 0$.

Remark 3.3 This is a reasonable assumption as the disturbances $f(x,t)$ and $d(t)$ have finite energy and hence are bounded, i.e., $f(x,t) \in \mathcal{L}_\infty([0, L])$ and $d(t) \in \mathcal{L}_\infty$.

Remark 3.4 The knowledge of the exact values for $f(x,t)$, $d(t)$, $\bar{f}$ and $\bar{d}$ are not required in the control design, which possesses stability robustness to variations of the unknown disturbances.

3.3 Control Design

We obtain the governing equations of the system as

$$\rho \ddot{w}(x,t) + EIw''''(x,t) - Tw''(x,t) = f(x,t), \tag{3.1}$$

$\forall (x,t) \in (0, L) \times [0, \infty)$. We obtain the boundary conditions of the system as

$$w'(0,t) = 0, \tag{3.2}$$
$$w''(L,t) = 0, \tag{3.3}$$
$$w(0,t) = 0, \tag{3.4}$$
$$-EIw'''(L,t) + Tw'(L,t) = u(t) + d(t) - M\ddot{w}(L,t), \tag{3.5}$$

$\forall t \in [0, \infty)$.

In order to suppress the vibration of the beam system governed by a nonhomogeneous PDE (3.1) under the unknown external disturbances $f(x,t)$ and $d(t)$, Lyapunov synthesis is used to construct a boundary control $u(t)$ at the right boundary of the flexible beam and analyze the closed-loop stability of the system.

3.3.1 Robust Boundary Control with a Disturbance Observer

When the system parameters EI, T, M are directly available, to stabilize the system given by governing (3.1) and boundary conditions (3.2)–(3.5), we propose the following boundary control

$$u(t) = -EIw'''(L,t) + Tw'(L,t) - M[k_1\dot{w}'(L,t) - k_2\dot{w}'''(L,t)] - ku_a(t) - \hat{d}(t), \quad (3.6)$$

where $\hat{d}(t)$ is the estimate of $d(t)$. k, k_1 and k_2 are the control gains and the auxiliary signal $u_a(t)$ is defined as

$$u_a(t) = \dot{w}(L,t) - k_2 w'''(L,t) + k_1 w'(L,t). \quad (3.7)$$

We define a vector $P(t)$ and the system parameter vector Φ as

$$P(t) = [w'''(L,t) \quad -w'(L,t) \quad k_1\dot{w}'(L,t) - k_2\dot{w}'''(L,t)], \quad (3.8)$$

$$\Phi = [EI \quad T \quad M]^T. \quad (3.9)$$

Then (3.6) can be rewritten in the following form

$$u(t) = -P(t)\Phi - ku_a(t) - \hat{d}(t). \quad (3.10)$$

Consider the following disturbance observer

$$\dot{\hat{d}}(t) = \gamma u_a(t) - \zeta_d \gamma \hat{d}(t), \quad (3.11)$$

where γ and ζ_d are two positive constants and the disturbance observer error is defined as

$$\tilde{d}(t) = d(t) - \hat{d}(t) \quad (3.12)$$

Since $\bar{d}$ is a positive constant, from (3.12) we have

$$\dot{\tilde{d}}(t) = -\gamma u_a(t) + \zeta_d \gamma \hat{d}(t). \quad (3.13)$$

Remark 3.5 For the proposed control (3.10), a disturbance observer and an auxiliary signal term are introduced to deal with the effect of unknown disturbances. The knowledge of the exact values for disturbances $f(x,t)$, $d(t)$ as well as their bounds $\bar{f}$ and $\bar{d}$ are not required, thus possessing stability robustness to variations with unknown disturbances is necessary.

Consider a Lyapunov function candidate

$$V_0(t) = V_1(t) + V_2(t) + \Delta(t) + \frac{1}{2}\gamma^{-1}\tilde{d}^2(t), \quad (3.14)$$

where the energy term $V_1(t)$ and an auxiliary term $V_2(t)$ and a small crossing term $\Delta(t)$ are defined as

$$V_1(t) = \frac{k_2\beta}{2}\rho\int_0^L [\dot{w}(x,t)]^2dx + \frac{k_2\beta}{2}EI\int_0^L [w''(x,t)]^2dx + \frac{k_2\beta}{2}T\int_0^L [w'(x,t)]^2dx, \quad (3.15)$$

$$V_2(t) = \frac{1}{2}Mu_a^2(t), \quad (3.16)$$

$$\Delta(t) = \alpha\rho\int_0^L x\dot{w}(x,t)w'(x,t)dx, \quad (3.17)$$

where α and β are two small positive weighting constants.

Lemma 3.1 *The Lyapunov function candidate given by (3.14) can be upper and lower bounded as*

$$0 \le \lambda_1(V_1(t) + V_2(t) + \tilde{d}^2(t)) \le V_0(t) \le \lambda_2(V_1(t) + V_2(t) + \tilde{d}^2(t)), \quad (3.18)$$

where λ_1 and λ_2 are two positive constants.

Proof

$$\begin{aligned}|\Delta(t)| &\le \alpha\rho L\int_0^L ([\dot{w}(x,t)]^2 + [w'(x,t)]^2)dx \\ &\le \alpha_1 V_1(t), \end{aligned} \quad (3.19)$$

where

$$\alpha_1 = \frac{2\alpha\rho L}{\min(k_2\beta\rho, k_2\beta T)}. \quad (3.20)$$

Then, we obtain

$$-\alpha_1 V_1(t) \le \Delta(t) \le \alpha_1 V_1(t). \quad (3.21)$$

Considering α is a small positive weighting constant satisfying $0 < \alpha < \frac{\min(\beta\rho, \beta T)}{2\rho L}$, we obtain

$$\alpha_2 = 1 - \alpha_1 = 1 - \frac{2\alpha\rho L}{\min(k_2\beta\rho, k_2\beta T)} > 0, \quad (3.22)$$

$$\alpha_3 = 1 + \alpha_1 = 1 + \frac{2\alpha\rho L}{\min(k_2\beta\rho, k_2\beta T)} > 1. \quad (3.23)$$

Then, we further have

$$0 \le \alpha_2 V_1(t) \le V_1(t) + \Delta(t) \le \alpha_3 V_1(t). \quad (3.24)$$

Given the Lyapunov function candidate (3.14), we obtain

$$\begin{aligned} 0 &\le \lambda_1(V_1(t) + V_2(t) + \tilde{d}^2(t)) \\ &\le V_1(t) + V_2(t) + \Delta(t) + \frac{1}{2}\gamma^{-1}\tilde{d}^2(t) \\ &\le \lambda_2(V_1(t) + V_2(t) + \tilde{d}^2(t)), \end{aligned} \tag{3.25}$$

where two positive constants λ_1 and λ_2 are given as

$$\lambda_1 = \min\left(\alpha_2, 1, \frac{1}{2\gamma}\right) = \min\left(\alpha_2, \frac{1}{2\gamma}\right), \tag{3.26}$$

$$\lambda_2 = \max\left(\alpha_3, 1, \frac{1}{2\gamma}\right) = \max\left(\alpha_3, \frac{1}{2\gamma}\right). \tag{3.27}$$

Lemma 3.2 *The time derivative of the Lyapunov function candidate (3.14) can be upper bounded with*

$$\dot{V}_0(t) \le -\lambda V_0(t) + \varepsilon_0, \tag{3.28}$$

where $\lambda > 0$ *and* $\varepsilon_0 > 0$.

Proof Differentiating (3.14) with respect to time leads to

$$\dot{V}_0(t) = \dot{V}_1(t) + \dot{V}_2(t) + \dot{\Delta}(t) + \gamma^{-1}\tilde{d}(t)\dot{\tilde{d}}(t). \tag{3.29}$$

The first term of the (3.29) is written as

$$\dot{V}_1(t) = A_1(t) + A_2(t) + A_3(t), \tag{3.30}$$

where

$$A_1(t) = k_2\beta\rho \int_0^L \dot{w}(x,t)\ddot{w}(x,t)dx, \tag{3.31}$$

$$A_2(t) = k_2\beta EI \int_0^L w''(x,t)\dot{w}''(x,t)dx, \tag{3.32}$$

$$A_3(t) = k_2\beta T \int_0^L w'(x,t)\dot{w}'(x,t)dx. \tag{3.33}$$

Substituting the governing equation (3.1) into A_1, we obtain

$$A_1(t) = k_2\beta \int_0^L \dot{w}(x,t)\left[-EIw''''(x,t) + Tw''(x,t) + f(x,t)\right]dx. \tag{3.34}$$

Using the boundary conditions and integrating (3.32) by parts, we obtain

$$\begin{aligned}
A_2(t) &= k_2\beta EI w''(x,t)\dot{w}'(x,t)\Big|_0^L - k_2\beta EI\int_0^L \dot{w}'(x,t)w'''(x,t)dx \\
&= -k_2\beta EI w'''(x,t)\dot{w}(x,t)\Big|_0^L + k_2\beta EI\int_0^L \dot{w}(x,t)w''''(x,t)dx \\
&= -k_2\beta EI w'''(L,t)\dot{w}(L,t) + k_2\beta EI\int_0^L \dot{w}(x,t)w''''(x,t)dx. \quad (3.35)
\end{aligned}$$

Using the boundary conditions and integrating (3.33) by part, we obtain

$$\begin{aligned}
A_3(t) &= k_2\beta T\int_0^L w'(x,t)d(\dot{w}(x,t)) \\
&= k_2\beta T w'(L,t)\dot{w}(L,t) - k_2\beta T\int_0^L \dot{w}(x,t)w''(x,t)dx. \quad (3.36)
\end{aligned}$$

Substituting (3.34), (3.35), and (3.36) into (3.30), we have

$$\dot{V}_1(t) = k_2\beta\left[-EI w'''(L,t) + T w'(L,t)\right]\dot{w}(L,t) + k_2\beta\int_0^L f(x,t)\dot{w}(x,t)dx. \quad (3.37)$$

Substituting (3.7) into (3.37), we obtain

$$\begin{aligned}
\dot{V}_1(t) &= \frac{\beta EI}{2}u_a^2(t) - \frac{\beta EI}{2}\{[\dot{w}(L,t)]^2 + k_2^2[w'''(L,t)]^2 + k_1^2[w'(L,t)]^2\} + \beta(k_2T - k_1EI)\dot{w}(L,t)w'(L,t) \\
&\quad + k_1k_2\beta EI w'(L,t)w'''(L,t) + k_2\beta\int_0^L f(x,t)\dot{w}(x,t)dx \quad (3.38) \\
\dot{V}_1(t) &\le \frac{\beta EI}{2}u_a^2(t) - \frac{\beta EI}{2}\{[\dot{w}(L,t)]^2 + k_2^2[w'''(L,t)]^2 + k_1^2[w'(L,t)]^2\} + \frac{\beta}{\delta_1}|k_2T - k_1EI|[\dot{w}(L,t)]^2 \\
&\quad + \beta\delta_1|k_2T - k_1EI|[w'(L,t)]^2 + k_1k_2\beta EI w'(L,t)w'''(L,t) + k_2\beta\delta_2\int_0^L[\dot{w}(x,t)]^2dx \\
&\quad + \frac{k_2\beta}{\delta_2}\int_0^L f^2(x,t)dx, \quad (3.39)
\end{aligned}$$

where δ_1 and δ_2 are two positive constants.

After differentiating the auxiliary signal (3.7), multiplying the resulting equation by M, and substituting (3.5), we obtain

$$\begin{aligned}
M\dot{u}_a(t) &= EI w'''(L,t) - T w'(L,t) + d(t) + M[k_1\dot{w}'(L,t) - k_2\dot{w}'''(L,t)] + u(t) \\
&= P(t)\Phi + d(t) + u(t). \quad (3.40)
\end{aligned}$$

Substituting (3.10) into (3.40), we have

$$M\dot{u}_a(t) = -ku_a(t) + d(t) - \hat{d}(t) \tag{3.41}$$

Substituting (3.41) into the second term of the (3.29), we have

$$\dot{V}_2(t) = Mu_a(t)\dot{u}_a(t) \tag{3.42}$$

The third term of the (3.29)

$$\begin{aligned}\dot{\Delta}(t) &= \alpha\rho\int_0^L (x\ddot{w}(x,t)w'(x,t) + x\dot{w}(x,t)\dot{w}'(x,t))dx \\ &= \alpha\int_0^L xw'(x,t)\left[-EIw''''(x,t) + Tw''(x,t) + f(x,t)\right]dx + \alpha\rho\int_0^L x\dot{w}(x,t)\dot{w}'(x,t)dx \\ &= B_1(t) + B_2(t) + B_3(t) + B_4(t),\end{aligned} \tag{3.43}$$

where

$$B_1(t) = -\alpha\int_0^L EIxw'(x,t)w''''(x,t)dx, \tag{3.44}$$

$$B_2(t) = \alpha\int_0^L Txw'(x,t)w''(x,t)dx, \tag{3.45}$$

$$B_3(t) = \alpha\int_0^L f(x,t)xw'(x,t)dx, \tag{3.46}$$

$$B_4(t) = \alpha\rho\int_0^L x\dot{w}(x,t)\dot{w}'(x,t)dx. \tag{3.47}$$

After integrating (3.44) by parts and using the boundary conditions, we obtain

$$\begin{aligned}B_1(t) = &-\alpha EILw'(L,t)w'''(L,t) + \alpha EI\int_0^L w'(x,t)w'''(x,t)dx \\ &+\alpha EI\int_0^L xw''(x,t)w'''(x,t)dx.\end{aligned} \tag{3.48}$$

Integrating the second term of (3.48) by parts, we have

$$\begin{aligned}B_1(t) = &-\alpha EILw'(L,t)w'''(L,t) - \alpha EI\int_0^L [w''(x,t)]^2dx \\ &+\alpha EI\int_0^L xw''(x,t)w'''(x,t)dx.\end{aligned} \tag{3.49}$$

Integrating the last term of (3.49) by parts, we have

$$B_1(t) = -\alpha EILw'(L,t)w'''(L,t) - \alpha EI\int_0^L [w''(x,t)]^2 dx$$
$$-\alpha EI\int_0^L \{[w''(x,t)]^2 + xw''(x,t)w'''(x,t)\}dx. \tag{3.50}$$

Combing the (3.49) and (3.50), we obtain

$$B_1(t) = -\alpha EILw'(L,t)w'''(L,t) - \frac{3\alpha EI}{2}\int_0^L [w''(x,t)]^2 dx. \tag{3.51}$$

After integrating (3.45) by parts and using the boundary conditions, we obtain

$$B_2(t) = \alpha TL[w'(L,t)]^2 - \alpha T\int_0^L \{[w'(x,t)]^2 + xw'(x,t)w''(x,t)\}dx. \tag{3.52}$$

Combining (3.45) and (3.52), we obtain

$$B_2(t) = \frac{\alpha TL}{2}[w'(L,t)]^2 - \frac{\alpha T}{2}\int_0^L [w'(x,t)]^2 dx. \tag{3.53}$$

Using Lemma 2.4, we have

$$B_2(t) \leq \frac{\alpha TL^2}{2}\int_0^L [w''(x,t)]^2 dx - \frac{\alpha T}{2}\int_0^L [w'(x,t)]^2 dx. \tag{3.54}$$

Using Lemma 2.2 to $B_3(t)$ term, we obtain

$$B_3(t) \leq \frac{\alpha L}{\delta_3}\int_0^L f^2(x,t)dx + \alpha L\delta_3\int_0^L [w'(x,t)]^2 dx, \tag{3.55}$$

where δ_3 is a positive constant. Integrating (3.47) by parts, we obtain

$$B_4(t) = \alpha\rho L[\dot{w}(L,t)]^2 - \alpha\rho\int_0^L \{[\dot{w}(x,t)]^2 + x\dot{w}(x,t)\dot{w}'(x,t)\}dx. \tag{3.56}$$

The last term in (3.56) is equal to $B_4(t)$, and we have

$$B_4(t) = \frac{\alpha\rho L}{2}[\dot{w}(L,t)]^2 - \frac{\alpha\rho}{2}\int_0^L [\dot{w}(x,t)]^2 dx. \tag{3.57}$$

Substituting (3.51), (3.54), (3.55) and (3.57) into (3.43) and using the boundary conditions, we obtain

$$\begin{aligned}\dot{\Delta}(t) \leq & -\alpha EILw'(L,t)w'''(L,t) - \frac{3\alpha EI}{2}\int_0^L [w''(x,t)]^2 dx + \frac{\alpha TL^2}{2}\int_0^L [w''(x,t)]^2 dx \\ & -\frac{\alpha T}{2}\int_0^L [w'(x,t)]^2 dx + \frac{\alpha L}{\delta_3}\int_0^L f^2(x,t)dx + \alpha L\delta_3 \int_0^L [w'(x,t)]^2 dx \\ & +\frac{\alpha\rho L}{2}[\dot{w}(L,t)]^2 - \frac{\alpha\rho}{2}\int_0^L [\dot{w}(x,t)]^2 dx. \end{aligned} \tag{3.58}$$

Substituting (3.39), (3.42) and (3.58) into (3.29), we obtain

$$\begin{aligned}\dot{V}_0(t) \leq & -\left(\frac{3\alpha EI}{2} - \frac{\alpha TL^2}{2}\right)\int_0^L [w''(x,t)]^2 dx - \left(\frac{\alpha T}{2} - \alpha L\delta_3\right)\int_0^L [w'(x,t)]^2 dx \\ & -\left(\frac{\alpha\rho}{2} - k_2\beta\delta_2\right)\int_0^L [\dot{w}(x,t)]^2 dx - \left(k - \frac{\beta EI}{2}\right)u_a^2(t) + (k_1k_2\beta - \alpha L)EIw'(L,t)w'''(L,t) \\ & +\frac{\alpha\rho L}{2}[\dot{w}(L,t)]^2 - \frac{\beta EI}{2}\{[\dot{w}(L,t)]^2 + k_2^2[w'''(L,t)]^2 + k_1^2[w'(L,t)]^2\} + \frac{\beta}{\delta_1}|k_2T - k_1EI|[\dot{w}(L,t)]^2 \\ & +\beta\delta_1|k_2T - k_1EI|[w'(L,t)]^2 + \left(\frac{k_2\beta}{\delta_2} + \frac{\alpha L}{\delta_3}\right)\int_0^L f^2(x,t)dx + \tilde{d}(t)u_a(t) \\ & +\gamma^{-1}\tilde{d}(t)\dot{\tilde{d}}(t). \end{aligned} \tag{3.59}$$

Substituting the disturbance observer (3.13) to $\dot{\tilde{d}}(t)$ term, we have

$$\begin{aligned}\dot{V}_0(t) \leq & -\left(\frac{3\alpha EI}{2} - \frac{\alpha TL^2}{2}\right)\int_0^L [w''(x,t)]^2 dx \\ & -\left(\frac{\alpha T}{2} - \alpha L\delta_3\right)\int_0^L [w'(x,t)]^2 dx - \left(\frac{\alpha\rho}{2} - k_2\beta\delta_2\right)\int_0^L [\dot{w}(x,t)]^2 dx - \left(k - \frac{\beta EI}{2}\right)u_a^2(t) \\ & -\left(\frac{\beta EI}{2} - \frac{\beta}{\delta_1}|k_2T - k_1EI| - \frac{\alpha\rho L}{2}\right)[\dot{w}(L,t)]^2 - \left(\frac{k_2^2\beta EI}{2} - \frac{EI}{\delta_4}|k_1k_2\beta - \alpha L|\right)[w'''(L,t)]^2 \\ & -\left[\frac{k_1^2\beta EI}{2} - \beta\delta_1|k_2T - k_1EI| - \delta_4 EI|k_1k_2\beta - \alpha L|\right][w'(L,t)]^2 \\ & +\left(\frac{k_2\beta}{\delta_2} + \frac{\alpha L}{\delta_3}\right)\int_0^L f^2(x,t)dx + \zeta_d\hat{d}(t)\tilde{d}(t). \end{aligned} \tag{3.60}$$

Since

$$\begin{aligned}\hat{d}(t)\tilde{d}(t) &= d(t)\tilde{d}(t) - \tilde{d}(t)^2 \\ &\leq \frac{1}{2}d(t)^2 + \frac{1}{2}\tilde{d}(t)^2 - \tilde{d}(t)^2 \\ &= \frac{1}{2}d(t)^2 - \frac{1}{2}\tilde{d}(t)^2, \end{aligned} \tag{3.61}$$

Substituting (3.61) into (3.60), we have

$$\begin{aligned}\dot{V}_0(t) &\leq -\lambda_3(V_1(t) + V_2(t)) - \frac{\zeta_d}{2}\tilde{d}^2(t) + \frac{\zeta_d}{2}d(t)^2 + \varepsilon_f \\ &\leq -\lambda_4(V_1(t) + V_2(t) + \tilde{d}^2(t)) + \varepsilon_0, \end{aligned} \tag{3.62}$$

where

$$\lambda_4 = \min\left(\lambda_3, \frac{\zeta_d}{2}\right), \tag{3.63}$$

$$\varepsilon_f = \left(\frac{k_2\beta}{\delta_2} + \frac{\alpha L}{\delta_3}\right)\int_0^L f^2 dx \leq \left(\frac{k_2\beta}{\delta_2} + \frac{\alpha L}{\delta_3}\right)\int_0^L \bar{f}^2 dx \in \mathcal{L}_\infty, \tag{3.64}$$

$$\varepsilon_0 = \varepsilon_f + \frac{\zeta_d}{2} d(t)^2, \tag{3.65}$$

and other constants $k, \alpha, \beta, \delta_1, \delta_2$ and δ_3 are chosen to satisfy the following conditions

$$\alpha < \frac{\min(k_2\beta\rho, k_2\beta T)}{2\rho L}, \tag{3.66}$$

$$\frac{\beta EI}{2} - \frac{\beta}{\delta_1}|k_2 T - k_1 EI| - \frac{\alpha\rho L}{2} \geq 0, \tag{3.67}$$

$$\frac{k_2^2\beta EI}{2} - \frac{EI}{\delta_4}|k_1 k_2 \beta - \alpha L| \geq 0, \tag{3.68}$$

$$\frac{k_1^2\beta EI}{2} - \beta\delta_1|k_2 T - k_1 EI| - \delta_4 EI|k_1 k_2\beta - \alpha L| \geq 0, \tag{3.69}$$

$$\sigma_1 = \frac{3\alpha EI}{2} - \frac{\alpha T L^2}{2} > 0, \tag{3.70}$$

$$\sigma_2 = \frac{\alpha T}{2} - \alpha L\delta_3 > 0, \tag{3.71}$$

$$\sigma_3 = \frac{\alpha\rho}{2} - k_2\beta\delta_2 > 0, \tag{3.72}$$

$$\sigma_4 = k - \frac{\beta EI}{2} > 0, \tag{3.73}$$

$$\lambda_3 = \min\left(\frac{2\sigma_1}{k_2\beta EI}, \frac{2\sigma_2}{k_2\beta T}, \frac{2\sigma_3}{k_2\beta\rho}, \frac{2\sigma_4}{M}\right) > 0. \tag{3.74}$$

Combining (3.25) and (3.62), we have

$$\dot{V}_0(t) \leq -\lambda V_0(t) + \varepsilon_0, \tag{3.75}$$

where $\lambda = \lambda_4/\lambda_2 > 0$.

With the above lemmas, we are ready to present the following stability theorem of the closed-loop beam system.

Theorem 3.1 *For the system dynamics described by (3.1) and boundary conditions (3.2) to (3.5), under Assumption 3.1, and the control (3.10), given that the initial conditions are bounded, we can conclude that the closed-loop system is*
(i) uniformly bounded: the state of the closed-loop system $w(x,t)$ will remain in the compact set Ω defined by

$$\Omega_1 := \{w(x,t) \in R| \mid w(x,t) \mid \leq D_1,\ \forall(x,t) \in [0,L]\times[0,\infty)\}, \tag{3.76}$$

where constant $D_1 = \sqrt{\frac{2L}{k_2\beta T\lambda_1}(V_0(0) + \frac{\varepsilon_0}{\lambda})}$,
(ii) uniformly ultimate bounded: the system state $w(x,t)$ *will eventually converge to the compact* Ω_2 *defined by*

$$\Omega_2 := \{w(x,t) \in R | \lim_{t\to\infty} | w(x,t) | \le D_2, \ \forall x \in [0,L]\}, \tag{3.77}$$

where constant $D_2 = \sqrt{\frac{2L\varepsilon_0}{k_2\beta T\lambda_1\lambda}}$.

Proof Multiplying (3.28) by $e^{\lambda t}$ yields

$$\frac{d}{dt}(V_0(t)e^{\lambda t}) \le \varepsilon_0 e^{\lambda t}. \tag{3.78}$$

Integrating of the above inequality, we obtain

$$V_0(t) \le \left(V_0(0) - \frac{\varepsilon_0}{\lambda}\right)e^{-\lambda t} + \frac{\varepsilon_0}{\lambda} \le V_0(0)e^{-\lambda t} + \frac{\varepsilon_0}{\lambda} \in \mathcal{L}_\infty, \tag{3.79}$$

which implies V_0 is bounded. Utilizing (2.7) and (3.15), we have

$$\frac{k_2\beta}{2L}Tw^2(x,t) \le \frac{k_2\beta}{2}T\int_0^L [w'(x,t)]^2 dx \le V_1(t) \le V_1(t) + V_2(t) \le \frac{1}{\lambda_1}V_0(t) \in \mathcal{L}_\infty. \tag{3.80}$$

Appropriately rearranging the terms of the above inequality, we obtain $w(x,t)$ is uniformly bounded as follows

$$|w(x,t)| \le \sqrt{\frac{2L}{k_2\beta T\lambda_1}\left(V_0(0)e^{-\lambda t} + \frac{\varepsilon_0}{\lambda}\right)} \le \sqrt{\frac{2L}{k_2\beta T\lambda_1}\left(V_0(0) + \frac{\varepsilon_0}{\lambda}\right)}, \ \forall (x,t) \in [0,L]\times[0,\infty). \tag{3.81}$$

Furthermore, from (3.81), we have

$$\lim_{t\to\infty} |w(x,t)| \le \sqrt{\frac{2L\varepsilon_0}{k_2\beta T\lambda_1\lambda}}, \ \forall x \in [0,L]. \tag{3.82}$$

Remark 3.6 In the above analysis, it is clear that the steady system state $w(x,t)$ can be made arbitrarily small provided that the design control parameters are appropriately selected. In practical applications, the design parameters should be adjusted carefully for achieving suitable transient performance and control action.

3.3.2 Adaptive Boundary Control with the System Parametric Uncertainties

When the system parameters EI, T, M are unknown, the following adaptive boundary control is designed to deal with the system parametric uncertainties

$$u(t) = -P(t)\hat{\Phi}(t) - ku_a(t) - \hat{d}(t), \tag{3.83}$$

where $\hat{\Phi}(t) = [\widehat{EI}(t) \quad \widehat{T}(t) \quad \widehat{M}(t)]^T$ is the parameter estimate vector. k is the control gain and the auxiliary signal $u_a(t)$ is defined as (3.7). We define the parameter estimate error vector $\tilde{\Phi}(t)$ as

$$\begin{aligned}\tilde{\Phi}(t) &= \Phi - \hat{\Phi}(t)\\ &= [EI - \widehat{EI}(t) \quad T - \widehat{T}(t) \quad M - \widehat{M}(t)]^T\\ &= [\widetilde{EI}(t) \quad \widetilde{T}(t) \quad \widetilde{M}(t)]^T.\end{aligned} \tag{3.84}$$

The adaptive law is designed as

$$\dot{\hat{\Phi}}(t) = \Gamma P^T(t)u_a(t) - \zeta_\Phi \Gamma \hat{\Phi}(t), \tag{3.85}$$

where $\Gamma \in R^{3\times 3}$ is a diagonal positive-definite matrix and ζ_Φ is a positive constant. The maximum and minimum eigenvalue of matrix Γ are given as λ_{max} and λ_{min} respectively. Since $\Phi = [EI \quad T \quad M]^T$ is a constant parameter vector, then from (3.84), we have

$$\dot{\tilde{\Phi}}(t) = -\Gamma P^T(t)u_a(t) + \zeta_\Phi \Gamma \hat{\Phi}(t). \tag{3.86}$$

Remark 3.7 For the proposed control (3.83), which is independent of system parameters, a parameter estimation term, a disturbance observer and an auxiliary signal term are introduced to compensate for the system parametric uncertainties and the effect of unknown disturbances.

Remark 3.8 All the signals in the boundary control can be measured by sensors or obtained by a backward difference algorithm. $w(L, t)$ can be sensed by a laser displacement sensor at the right boundary of the beam, $w'(L, t)$ can be measured by an inclinometer and $w'''(L, t)$ can be obtained by a shear force sensor. In practice, the effect of measurement noise from sensors is unavoidable, which will affect the control implementation, especially when the high order differentiating terms with respect to time exist. In our proposed control (3.10) and (3.83), $\dot{w}(L, t)$ and $\dot{w}'''(L, t)$ with only one time differentiating with respect to time can be calculated with a backward difference algorithm. It is noted that differentiating twice and three times position $w(L, t)$ with respect to time to get $\ddot{w}(L, t)$ and $\dddot{w}(L, t)$ respectively, are undesirable in practice due to noise amplification. For these cases, observers are needed to design to estimate the state's values according to the boundary conditions.

Remark 3.9 Both control techniques (3.10) and (3.83) are based on the distributed parameter model (3.1)–(3.5), and the spillover problems associated with traditional truncated model-based approaches caused by ignoring high-frequency modes are avoided.

Consider the Lyapunov function candidate

$$V(t) = V_0(t) + \frac{1}{2}\tilde{\Phi}^T(t)\Gamma^{-1}\tilde{\Phi}(t). \tag{3.87}$$

Lemma 3.3 *The Lyapunov function candidate given by (3.87), can be upper and lower bounded as*

$$0 \le \lambda_{1a}(V_1(t) + V_2(t) + ||\tilde{\Phi}(t)||^2 + \tilde{d}^2(t)) \le V(t) \le \lambda_{2a}(V_1(t) + V_2(t) + ||\tilde{\Phi}(t)||^2 + \tilde{d}^2(t)), \tag{3.88}$$

where λ_{1a} and λ_{2a} are two positive constants.

Proof From Lemma 2.3, we have

$$\frac{1}{2\lambda_{max}}||\tilde{\Phi}(t)||^2 \le \frac{1}{2}\tilde{\Phi}^T(t)\Gamma^{-1}\tilde{\Phi}(t) \le \frac{1}{2\lambda_{min}}||\tilde{\Phi}(t)||^2. \tag{3.89}$$

Combining (3.25), (3.89) and (3.87), we have

$$0 \le \lambda_{1a}(V_1(t) + V_2(t) + ||\tilde{\Phi}(t)||^2 + \tilde{d}^2(t)) \le V(t) \le \lambda_{2a}(V_1(t) + V_2(t) + ||\tilde{\Phi}(t)||^2 + \tilde{d}^2(t)), \tag{3.90}$$

where $\lambda_{1a} = \min(\lambda_1, \frac{1}{2\lambda_{max}})$ and $\lambda_{2a} = \max(\lambda_2, \frac{1}{2\lambda_{min}})$ are two positive constants.

Lemma 3.4 *The time derivative of the Lyapunov function candidate (3.87) can be upper bounded with*

$$\dot{V}(t) \le -\lambda_a V(t) + \varepsilon, \tag{3.91}$$

where $\lambda_a > 0$ and $\varepsilon > 0$.

Proof Differentiating (3.87) with respect to time leads to

$$\dot{V}(t) = \dot{V}_0(t) + \tilde{\Phi}^T(t)\Gamma^{-1}\dot{\tilde{\Phi}}(t). \tag{3.92}$$

Substituting (3.83) into (3.40), we have

$$M\dot{u}_a(t) = P(t)\Phi - P(t)\hat{\Phi}(t) - ku_a(t) + d(t) - \hat{d}(t) \tag{3.93}$$

Substituting (3.93) into the second term of the (3.29), we have

$$\dot{V}_2(t) = M u_a(t)\dot{u}_a(t) \tag{3.94}$$

Substituting (3.39), (3.94) and (3.58) into (3.92), we obtain

$$\begin{aligned}
\dot{V}(t) \leq & -\left(\frac{3\alpha EI}{2} - \frac{\alpha T L^2}{2}\right)\int_0^L [w''(x,t)]^2 dx \\
& -\left(\frac{\alpha T}{2} - \alpha L \delta_3\right)\int_0^L [w'(x,t)]^2 dx - \left(\frac{\alpha\rho}{2} - k_2\beta\delta_2\right)\int_0^L [\dot{w}(x,t)]^2 dx - \left(k - \frac{\beta EI}{2}\right)u_a^2(t) \\
& -\left(\frac{\beta EI}{2} - \frac{\beta}{\delta_1}|k_2 T - k_1 EI| - \frac{\alpha\rho L}{2}\right)[\dot{w}(L,t)]^2 - \left(\frac{k_2^2\beta EI}{2} - \frac{EI}{\delta_4}|k_1 k_2 \beta - \alpha L|\right)[w'''(L,t)]^2 \\
& -\left[\frac{k_1^2\beta EI}{2} - \beta\delta_1|k_2 T - k_1 EI| - \delta_4 EI|k_1 k_2\beta - \alpha L|\right][w'(L,t)]^2 \\
& +\left(\frac{k_2\beta}{\delta_2} + \frac{\alpha L}{\delta_3}\right)\int_0^L f^2(x,t)dx + \zeta_a\hat{d}(t)\tilde{d}(t) + \tilde{\Phi}^T(t)\Gamma^{-1}\dot{\hat{\Phi}}(t) + P(t)\tilde{\Phi}(t)u_a(t).
\end{aligned} \tag{3.95}$$

Substituting the adaptive law (3.86) into (3.95), we have

$$\begin{aligned}
\dot{V}(t) &\leq -\lambda_4\left[V_1(t) + V_2(t) + \tilde{d}^2(t)\right] + \zeta_\Phi\tilde{\Phi}^T(t)\hat{\Phi}(t) + \varepsilon_0 \\
&\leq -\lambda_4\left[V_1(t) + V_2(t) + \tilde{d}^2(t)\right] - \frac{\zeta_\Phi}{2}||\tilde{\Phi}(t)||^2 + \frac{\zeta_\Phi}{2}||\Phi||^2 + \varepsilon_0 \\
&\leq -\lambda_{4a}\left[V_1(t) + V_2(t) + \tilde{d}^2(t) + ||\tilde{\Phi}(t)||^2\right] + \frac{\zeta_\Phi}{2}||\Phi||^2 + \varepsilon_0,
\end{aligned} \tag{3.96}$$

where $\lambda_{4a} = \min(\lambda_4, \ \frac{\zeta_\Phi}{2})$ is a positive constant. Combining (3.90) and (3.96), we have

$$\dot{V}(t) \leq -\lambda_a V(t) + \varepsilon, \tag{3.97}$$

where $\lambda_a = \lambda_{4a}/\lambda_{2a} > 0$ and $\varepsilon = \varepsilon_0 + \frac{\zeta_\Phi}{2}||\Phi||^2 > 0$.

With the above lemmas, we are ready to present the following stability theorem of the closed-loop beam system.

Theorem 3.2 *For the system dynamics described by (3.1) and boundary conditions (3.2) to (3.5), under Assumption 3.1, and the control (3.83), given that the initial conditions are bounded, we can conclude that*
(i) uniform boundedness (UB): the state of the closed-loop system $w(x,t)$ will remain in the compact set Ω defined by

$$\Omega_3 := \{w(x,t) \in R| \ | w(x,t) | \leq D_3, \ \forall (x,t) \in [0, L] \times [0, \infty)\}, \tag{3.98}$$

where constant $D_3 = \sqrt{\frac{2L}{k_2\beta T\lambda_{1a}}(V(0) + \frac{\varepsilon}{\lambda_a})}$,
(ii) uniform ultimate boundedness (UUB): the system state $w(x,t)$ will eventually

converge to the compact Ω_4 defined by

$$\Omega_4 := \{w(x,t) \in R | \lim_{t\to\infty} | w(x,t) | \le D_4, \ \forall x \in [0, L]\}, \tag{3.99}$$

where constant $D_4 = \sqrt{\frac{2L\varepsilon}{k_2 \beta T \lambda_{1a} \lambda_a}}$.

Proof Multiplying (3.91) by $e^{\lambda_a t}$ yields

$$\frac{d}{dt}(V(t)e^{\lambda_a t}) \le \varepsilon e^{\lambda_a t}. \tag{3.100}$$

Integrating of the above inequality, we obtain

$$V(t) \le \left(V(0) - \frac{\varepsilon}{\lambda_a}\right) e^{-\lambda_a t} + \frac{\varepsilon}{\lambda_a} \le V(0)e^{-\lambda_a t} + \frac{\varepsilon}{\lambda_a} \in \mathcal{L}_\infty, \tag{3.101}$$

which implies V is bounded. Utilizing (3.15), we have

$$\frac{k_2\beta}{2L} T w^2(x,t) \le \frac{k_2\beta}{2} T \int_0^L [w'(x,t)]^2 dx \le V_1(t) \le V_1(t) + V_2(t) \le \frac{1}{\lambda_{1a}} V(t) \in \mathcal{L}_\infty. \tag{3.102}$$

Appropriately rearranging the terms of the above inequality, we obtain $w(x,t)$ is uniformly bounded as follows

$$|w(x,t)| \le \sqrt{\frac{2L}{k_2\beta T \lambda_{1a}} \left(V(0)e^{-\lambda_a t} + \frac{\varepsilon}{\lambda_a}\right)} \le \sqrt{\frac{2L}{k_2\beta T \lambda_{1a}} \left(V(0) + \frac{\varepsilon}{\lambda_a}\right)}, \ \forall (x,t) \in [0, L] \times [0, \infty). \tag{3.103}$$

Furthermore, from (3.103), we have

$$\lim_{t\to\infty} |w(x,t)| \le \sqrt{\frac{2L\varepsilon}{k_2\beta T \lambda_{1a}\lambda_a}}, \ \forall x \in [0, L]. \tag{3.104}$$

Remark 3.10 From the similar analysis of Remark 3.6, we can conclude that the steady system state $w(x,t)$ with proposed adaptive boundary control can be made arbitrarily small by choosing control gains in (3.83) appropriately.

Remark 3.11 From (3.90), we can obtain the parameter estimation error $\tilde{\Phi}(t)$ are bounded $\forall t \in [0, \infty)$.

Table 3.1 Parameters of the beam

Parameter	Description	Value
L	Length of beam	1 m
EI	Bending stiffness of the beam	5 Nm^2
T	Tension	6 N
ρ	Mass per unit length of the beam	1 kg/m
M	Mass of the tip payload	5 kg

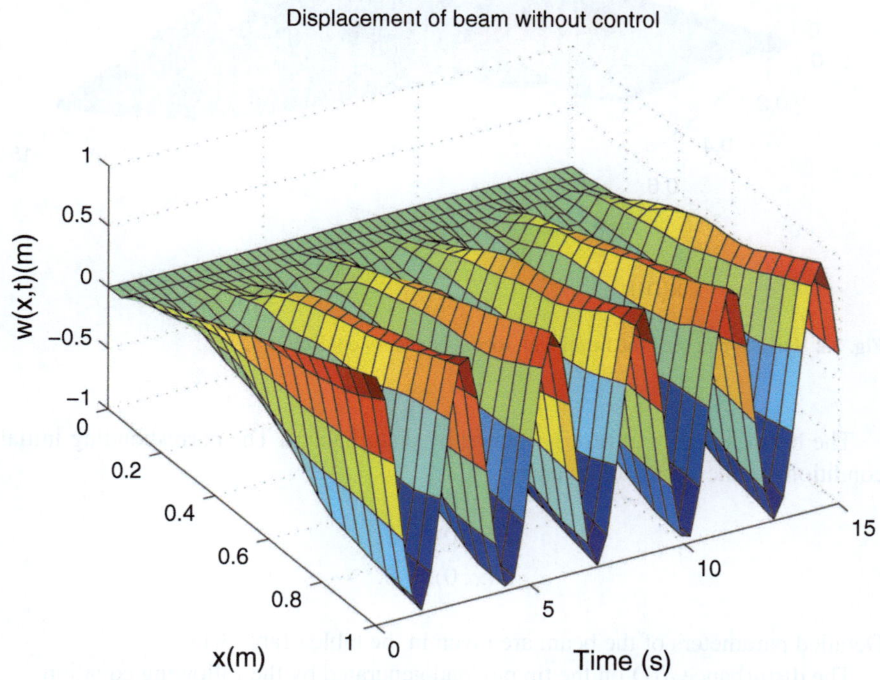

Fig. 3.1 Displacement of the beam without control

3.4 Numerical Simulations

Simulations for a beam of length 1 m under unknown external disturbances are carried out to demonstrate the effectiveness of the proposed boundary vibration control (3.10) and (3.83). Due to the unknown term $f(x, t)$, the eigenfunction expansion method based on Fourier series in [8] cannot be used for the beam system (3.1)–(3.5). In this chapter, we select the finite difference scheme to simulate the system performance with the proposed boundary control.

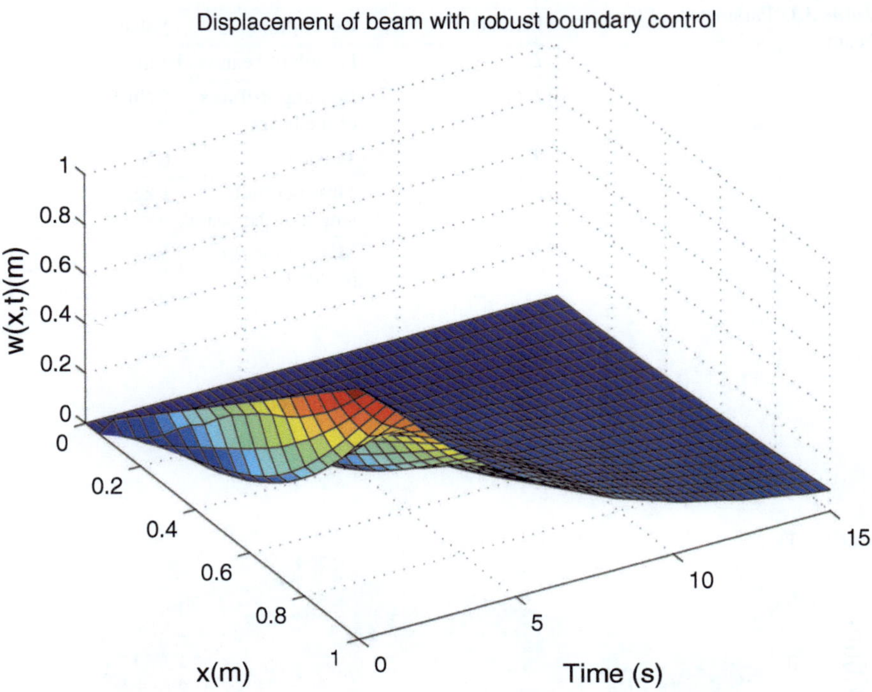

Fig. 3.2 Displacement of the beam with robust boundary control

The beam, initially at rest, is excited by disturbances. The corresponding initial conditions of the beam are given as

$$w(x,0) = x,$$
$$\dot{w}(x,0) = 0.$$

Detailed parameters of the beam are given in the table (Table 3.1).

The disturbance $d(t)$ on the tip payload generated by the following equation

$$d(t) = 0.1 + 0.1\sin(\pi t) + 0.1\sin(2\pi t) + 0.1\sin(3\pi t). \tag{3.105}$$

The distributed disturbance $f(x,t)$ along the beam is described as

$$f(x,t) = \frac{[1 + \sin(0.1\pi xt) + \sin(0.2\pi xt) + \sin(0.3\pi xt)]x}{20L}. \tag{3.106}$$

Displacement of the beam for free vibration, i.e., $u(t) = 0$, under the disturbances $f(x,t)$ and $d(t)$ is shown in Fig. 3.1. Displacement of the beam with the proposed robust boundary control (3.10) as well as the disturbance observer (3.11), by choosing $k = 10$, $k_1 = 0.04$, $k_2 = 0.02$, $\zeta_d = 0.001$ and $\gamma = 1$, is shown in Fig. 3.2. When

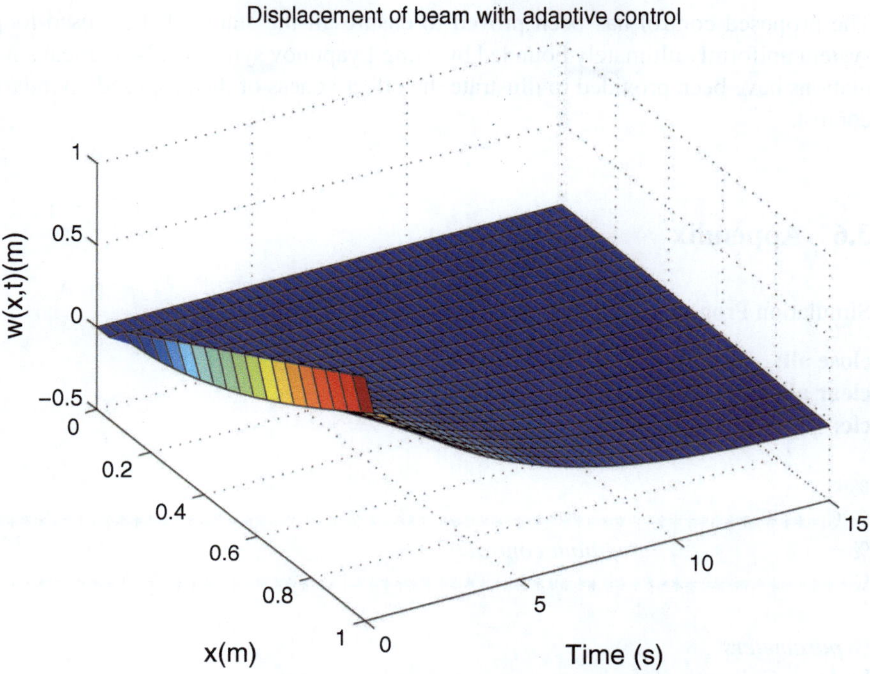

Fig. 3.3 Displacement of the beam with adaptive boundary control

there are system parametric uncertainties and disturbances uncertainties, displacement of the beam with the proposed adaptive boundary control (3.83), by choosing $k = 2$, $\zeta_\Phi = \zeta_d = 0.02$, and $\gamma = 1$, $\Gamma = \text{diag}\{1, 1, 1\}$, is shown in Fig. 3.3. Figures 3.2 and 3.3 illustrate that both robust boundary control (3.10) and adaptive boundary control (3.83) are able to stabilize the beam at the small neighborhood of its equilibrium position.

3.5 Conclusion

In this chapter, two cases for the vibrating beam system under unknown spatiotemporally varying distributed disturbance $f(x, t)$ and unknown time-varying boundary disturbance $d(t)$ have been investigated: (i) robust boundary control for disturbances uncertainties, and (ii) adaptive boundary control for both the system parametric uncertainties and disturbances uncertainties. Firstly, a boundary control which can achieve the suppression of the vibrating beam system coupled with a disturbance observer which can attenuate the unknown external disturbances has been introduced for the beam system. In the second case, an adaptive boundary control which can compensate both the parametric uncertainties and disturbances uncertainties has been studied.

The proposed control has been proved to ensure all the states of the closed-loop system uniformly ultimately bounded by using Lyapunov synthesis. Numerical simulations have been provided to illustrate the effectiveness of the proposed boundary control.

3.6 Appendix

Simulation Program:

```
close all;
clear all;
clc;

%%
% %**********************************************************
%                     without control
%**********************************************************

% parameters
L=1;      % length of the beam
EI=5;     % bending stiffness of the beam
T=6;      % tension
rho=1;    % mass per unit length of the beam
Ms=5;     % mass of the tip payload


% constant parameters
nt=80000; % length of time domain
nx=20;    % length of space domain
tf=15;    % time of simulation
Ttr=50;   % sampling for drawing

% compute the mesh spacing and time step
dx=L/nx;     % spacing step
dt=tf/nt;    % time step


% drawing coordinate
x = linspace(0,L,nx+1);          % create spaceline for drawing
t_tr = linspace(0,tf,Ttr+1);  % create timeline for drawing

% create matrix to save data contains
w=zeros(nx+1,nt+1);              % vibrations at position x for time t
```

```
w3D_free=zeros(Ttr+1,nx+1); % save data from w
d=zeros(nt+1,1);            % boundary disturbance
f=zeros(nx+1,nt+1);         % distributed disturbance

% boundary disturbance
for j=1:nt+1
    d(j) = 0.1 + 0.1*sin(pi*(j−1)*dt) + 0.1*sin(2*pi*(j−1)*dt) + 0.1*
        sin(3*pi*(j−1)*dt);
end

% Distributed disturbance
for i=1:nx+1
   for j=1:nt+1
       f(i,j)=(1 + sin(0.1*pi*(i−1)*dx*(j−1)*dt) + sin(0.2*pi*(i−1)*dx*
           (j−1)*dt)...
       + sin(0.3*pi*(i−1)*dx*(j−1)*dt) )*(i−1)*dx/(20*L);
   end
end

% initial condition
for i=1:nx+1
    w(i,1)=(i−1)*dx;
    w(i,2)=w(i,1);
end
w(1,:)=0;
w(2,:)=w(1,:);

w3D_free(1,:)=w(:,1);
w3D_free(2,:)=w(:,2);

% main cycle
for j=3:nt+1

    for i=3:nx−1
        wxx=( w(i+1,j−1)−2*w(i,j−1)+w(i−1,j−1) )/dx^2;
        wxxxx=( w(i+2,j−1)−4*w(i+1,j−1)+6*w(i,j−1)−4*w(i−1,j−1)+w(i−2,
            j−1) )/dx^4;
        w(i,j)=2*w(i,j−1)−w(i,j−2)+(T*wxx−EI*wxxxx+f(i,j))*dt^2/rho;
    end

    wxl=( w(nx+1,j−1)−w(nx,j−1) )/dx;
    wxxxl=( w(nx+1,j−1)−3*w(nx,j−1)+3*w(nx−1,j−1)−w(nx−2,j−1) )/dx^3;

    % boundary condition at L
```

```
    w(nx+1,j)=2*w(nx+1,j-1) - w(nx+1,j-2) - + (d(j-1)+EI*wxxxl-T*wxl)*
        dt^2/M;
    w(nx,j)=(w(nx+1,j)+w(nx-1,j))/2;

    % saving data for drawing
    if mod(n-1,(nt/Ttr))==0
        w3D_free(1+(j-1)*Ttr/nt,:)=w(:,j);
    end

end

% make a draw
figure(1);
surf(x,t_tr,w3D_free);
title('Displacement of beam without control');
ylabel('Time(s)','Fontsize',12);
xlabel('x(m)','Fontsize',12);
zlabel('w(x,t)(m)','Fontsize',12);
view([60 45]);


%%
%*************************************************************
%                    Exact model control
%*************************************************************

K=10;           % control gain
k_1=0.04;       % control gaingama=1;
k_2=0.02;       % control gain
zeta_d=0.001;   % disturbance observer parameter
gama=1;         % disturbance observer parameter

% create matrix to save data contains
w=zeros(nx+1,nt+1);                  % vibrations at position x for time t
w3D_model=zeros(Ttr+1,nx+1);         % save data from w
u_model=zeros(nt+1,1);               % control input
u3D_model=zeros(Ttr+1,1);            % save data from u_model
ua=zeros(nt+1,1);                    % auxiliary signal
d_estimate=zeros(nt+1,1);            % disturbance observer

% initial condition
dbar=max(abs(d));
d_estimate(1:2)= dbar;
```

```
for i=1:nx+1
    w(i,1)=(i−1)∗dx;
    w(i,2)=w(i,1);
end
w(1,:)=0;
w(2,:)=w(1,:);

w3D_model(1,:)=w(:,1);
w3D_model(2,:)=w(:,2);

% main cycle
for j=3:nt+1

    for i=3:nx−1
        wxx=(w(i+1,j−1)−2∗w(i,j−1)+w(i−1,j−1))/dx^2;
        wxxxx=(w(i+2,j−1)−4∗w(i+1,j−1)+6∗w(i,j−1)−4∗w(i−1,j−1)+
            w(i−2,j−1))/dx^4;
        w(i,j)=2∗w(i,j−1)−w(i,j−2)+(T∗wxx−EI∗wxxxx+f(i,j))∗dt^2/rho;
    end

    % controller design
    wxl=(w(nx+1,j−1)−w(nx,j−1))/dx;
    wxxxl=(w(nx+1,j−1)−3∗w(nx,j−1)+3∗w(nx−1,j−1)−w(nx−2,j−1))/dx^3;
    wxxxtl=(w(nx+1,j−1)−3∗w(nx,j−1)+3∗w(nx−1,j−1)−w(nx−2,j−1)
        −w(nx+1,j−2)+3∗w(nx,j−2)−3∗w(nx−1,j−2)+w(nx−2,j−2))/(dt∗dx^3);
    wxtl=(w(nx+1,j−1)−w(nx,j−1)−w(nx+1,j−2)+w(nx,j−2))/(dx∗dt);
    wtl=(w(nx+1,j−1)−w(nx+1,j−2))/dt;

    ua(j)=wtl+k_1∗wxl−k_2∗wxxxl;
    d_estimate(j)= (dt∗gama∗ua(j)+ d_estimate(j−1)) /(1+gama∗dt∗zeta_d);

    u_model(j)= −EI∗wxxxl+T∗wxl−M∗(k_1∗wxtl−k_2∗wxxxtl)−k∗ua(j)
        −d_estimate(j);

    % boundary condition at L
    w(nx+1,j)=2∗w(nx+1,j−1) − w(nx+1,j−2) + (u_model(j−1)+d(j−1)+
        EI∗wxxxl−T∗wxl)∗dt^2/M;
    w(nx,j)=(w(nx+1,j)+w(nx−1,j))/2;

    % saving data for drawing
    if mod(j−1,(nt/Ttr))==0
        w3D_model(1+(j−1)∗Ttr/nt,:)=w(:,j);
        u3D_model(1+(j−1)∗Ttr/nt)=u_model(j);
    end
end
```

```
% drawing coordinate
x=linspace(0,L,nx+1);     % create spaceline for drawing
t_tr=linspace(0,tf,Ttr+1);  % create timeline for drawing

% make a draw
figure(2);
surf(x,t_tr,w3D_model);
title('Displacement of beam with robust boundary control');
ylabel('Time (s)','Fontsize',12);
xlabel('x(m)','Fontsize',12);
zlabel('w(x,t)(m)','Fontsize',12);
view([60 45]);
```

References

1. G. Zhu, S.S. Ge, A quasi-tracking approach for finite-time control of a mass-beam system. Automatica **34**(7), 881–888 (1998)
2. A. Smyshlyaev, B. Guo, M. Krstic, Arbitrary decay rate for Euler-Bernoulli beam by backstepping boundary feedback. IEEE Trans. Autom. Control **54**(5), 1134–1140 (2009)
3. B. Guo, K. Yang, Dynamic stabilization of an Euler-Bernoulli beam equation with time delay in boundary observation. Automatica **45**(6), 1468–1475 (2009)
4. S.S. Ge, T.H. Lee, G. Zhu, Improving regulation of a single-link flexible manipulator with strain feedback. IEEE Trans. Robot. Autom. **14**(1), 179–185 (1998)
5. Q. Hu, Adaptive output feedback sliding-mode manoeuvring and vibration control of flexible spacecraft with input saturation. IET Control Theory Appl. **2**(6), 467–478 (2008)
6. S.S. Ge, W. He, B.V.E. How, Y.S. Choo, Boundary control of a coupled nonlinear flexible marine riser. IEEE Trans. Control Syst. Technol. **18**(5), 1080–1091 (2010)
7. J.-Y. Choi, K.-S. Hong, K.-J. Yang, Exponential stabilization of an axially moving tensioned strip by passive damping and boundary control. Modal Anal. **10**(5), 661–682 (2004)
8. J.D. Logan, *Applied Mathematics*, 3rd edn. (Wiley, New York, 2006)
9. M. Balas, Feedback control of flexible systems. IEEE Trans. Autom. Control **23**(4), 673–679 (1978)
10. M.W. Vandegrift, F.L. Lewis, S.Q. Zhu, Flexible-link robot arm control by a feedback linearization/singular perturbation approach. J. Robot. Syst. **11**(7), 591–603 (1994)
11. S.S. Ge, T.H. Lee, G. Zhu, A nonlinear feedback controller for a single-link flexible manipulator based on a finite element model. J. Robot. Syst. **14**(3), 165–178 (1997)
12. M.J. Balas, Active control of flexible systems. J. Optim. Theory Appl. **25**, 415–436 (1978)
13. L. Meirovitch, H. Baruh, On the problem of observation spillover in self-adjoint distributed systems. J. Optim. Theory Appl. **39**(2), 269–291 (1983)
14. B. How, S. Ge, Y. Choo, Active control of flexible marine risers. J. Sound Vib. **320**(4), 758–776 (2009)
15. O. Morgul, Dynamic boundary control of the Timoshenko beam. Automatica **28**(6), 1255–1260 (1992)
16. K. Endo, F. Matsuno, H. Kawasaki, Simple boundary cooperative control of two one-link flexible arms for grasping. IEEE Trans. Autom. Control **54**(10), 2470–2476 (2009)
17. M. Krstic, A. Smyshlyaev, *Boundary Control of PDEs: A Course on Backstepping Designs* (Society for Industrial and Applied Mathematics, Philadelphia, 2008)

18. K.-J. Yang, K.-S. Hong, F. Matsuno, Robust boundary control of an axially moving string by using a PR transfer function. IEEE Trans. Autom. Control **50**(12), 2053–2058 (2005)
19. K.-J. Yang, K.-S. Hong, F. Matsuno, Energy-based control of axially translating beams: varying tension, varying speed, and disturbance adaptation. IEEE Trans. Control Syst. Technol. **13**(6), 1045–1054 (2005)
20. Q.C. Nguyen, K.-S. Hong, Asymptotic stabilization of a nonlinear axially moving string by adaptive boundary control. Journal of Sound and Vibration **329**(22), 4588–4603 (2010)
21. K. Li, T. Leung, Y. Hu, Sliding mode control of distributed parameter systems. Automatica **30**(12), 1961–1966 (1994)
22. S.S. Ge, T.H. Lee, G. Zhu, F. Hong, Variable structure control of a distributed-parameter flexible beam. J. Robot. Syst. **18**(1), 17–27 (2001)
23. S.S. Ge, T.H. Lee, G. Zhu, Energy-based robust controller design for multi-link flexible robots. Mechatronics **6**(7), 779–798 (1996)
24. S.S. Ge, T.H. Lee, Z. Wang, Model-free regulation of multi-link smart materials robots. IEEE/ASME Trans. Mechatron. **6**(3), 346–351 (2001)
25. K.-S. Hong, J. Bentsman, Application of averaging method for integro-differential equations to model reference adaptive control of parabolic systems. Automatica **30**(9), 1415–1419 (1994)
26. B. Bamieh, F. Paganini, M.A. Dahleh, Distributed control of spatially invariant systems. IEEE Trans. Autom. Control **47**(7), 1091–1107 (2002)
27. F. Wu, Distributed control for interconnected linear parameter-dependent systems. IEE Proc. Control Theory Appl. **150**(5), 518–527 (2003)
28. H. Banks, R. Smith, Y. Wang, *Smart Material Structures: Modeling, Estimation, and Control* (Wiley, New York, 1997)
29. M. Fard, S. Sagatun, Exponential stabilization of a transversely vibrating beam via boundary control. J. Sound Vib. **240**(4), 613–622 (2001)
30. A. Baz, Dynamic boundary control of beams using active constrained layer damping. Mech. Syst. Signal Process. **11**(6), 811–825 (1997)
31. M.P. Fard, S.I. Sagatun, Exponential stabilization of a transversely vibrating beam by boundary control via Lyapunov direct method. J. Dyn. Syst. Meas. Control **123**(2), 195–200 (2001)
32. T.D. Nguyen, Boundary output feedback of second-order distributed parameter systems. Syst. Control Lett. **58**(7), 519–528 (2009)
33. K.D. Do, J. Pan, Boundary control of transverse motion of marine risers with actuator dynamics. J. Sound Vib. **318**(4), 768–791 (2008)
34. Y. Song, X. He, Z. Liu, W. He, C. Sun, F.-Y. Wang, Parallel control of distributed parameter systems. IEEE Trans. Cybern. (2018). https://doi.org/10.1109/TCYB.2018.2849569 (In Press)
35. W. He, S.S. Ge, Vibration control of a flexible beam with output constraint. IEEE Trans. Ind. Electron. **62**(8), 5023–5030 (2015)
36. W. He, X. He, S.S. Ge, Vibration control of flexible marine riser systems with input saturation. IEEE/ASME Trans. Mechatron. **21**(1), 254–265 (2016)
37. W. He, C. Sun, S.S. Ge, Top tension control of a flexible marine riser by using integral-barrier Lyapunov function. IEEE/ASME Trans. Mechatron. **2**(20), 497–505 (2015)
38. W. He, S.S. Ge, B.V.E. How, Y.S. Choo, K.-S. Hong, Robust adaptive boundary control of a flexible marine riser with vessel dynamics. Automatica **47**(4), 722–732 (2011)

Chapter 4
Vibration Control of a Flexible Beam with Output Constraint

4.1 Introduction

Lyapunov theory, one of the most successfully and widely used tools, provides a means of determining stability without explicit knowledge of system solutions. Many remarkable results [1–4] have been presented for flexible systems based on Lyapunov's direct method. Barrier Lyapunov function is a novel concept that can be employed to deal with control problems with output constraints [5–7]. In [5], a barrier Lyapunov function is employed for control of SISO nonlinear systems with an output constraint. A novel asymmetric time-varying barrier Lyapunov function is used in [7] to ensure the time-varying output constraint satisfaction for strict feedback nonlinear systems. In the neural control field, two challenging and open problems are addressed in [6] by using a barrier Lyapunov function in the presence of unknown functions. However, in all the papers mentioned above, barrier Lyapunov functions are designed for linear or nonlinear ODE systems. There is little information about how to handle the constraints for PDEs and there is a need to explore an effective method for the control of flexible systems with constraint problems.

For control of DPS modeled by PDEs, one of the effective control methods is based on the discretization of the PDE into a finite number of ODEs [8, 9]. The other method is to design a control based on the original PDEs without any model discretization [10–14]. In this chapter, the problem for control of an Euler-Bernoulli beam with boundary output constraint is addressed. We design a boundary barrier control scheme and apply the proposed Lyapunov function to the original PDEs to avoid the spillover problem. A novel form of Lyapunov function combining both the integral Lyapunov function and the barrier Lyapunov function is employed for the control design and stability analysis of the system.

The rest chapter is organized as follows. Based on the dynamics model of Euler-Bernoulli beam, a model-based barrier control and an adaptive barrier control are designed in Sect. 4.2. The numerical simulations are illustrated to verify the control effectiveness of the controllers in Sect. 4.3. The conclusions are presented in Sect. 4.4.

W. He and J. Liu, *Active Vibration Control and Stability Analysis of Flexible Beam Systems*, https://doi.org/10.1007/978-981-10-7539-1_4

4.2 Control Design

The control objectives of this chapter are as follows. (i) to suppress the transverse vibrations $w(x,t)$ of the beam; (ii) to keep the boundary output $w(L,t)$ remain in the constrained space, i.e. to guarantee $w(L,t)$ satisfying $| w(L,t) | < l_0$, where l_0 is a positive constant. Figure 4.1 shows a typical beam-based structure. The left boundary of the beam is fixed at the origin.

In this chapter, the flexible beam system with flexural rigidity is modeled as the Euler-Bernoulli beam structure since neglecting the rotary inertia of the beam and the diameter-to-length of the riser is small. Consider the following Euler-Bernoulli beam system [15] with a boundary disturbance, where the governing equation is given as

$$\rho\ddot{w}(x,t) + EIw''''(x,t) - Tw''(x,t) = 0, \quad \forall (x,t) \in (0,L) \times [0,\infty) \tag{4.1}$$

and the boundary conditions as

$$w'(0,t) = w''(L,t) = w(0,t) = 0, \tag{4.2}$$

$$-EIw'''(L,t) + Tw'(L,t) = u(t) - M\ddot{w}(L,t) + d(t), \quad \forall t \in [0,\infty) \tag{4.3}$$

where x and t represent the independent spatial and time variables respectively.

Assumption 4.1 For the unknown boundary disturbance $d(t)$, we assume that there exists a constant $\bar{d} \in \mathbb{R}^+$, such that $|d(t)| \leq \bar{d}$, $\forall t \in [0,\infty)$. This is a reasonable assumption for the time-varying $d(t)$ has finite energy and thus is bounded, i.e., $d(t) \in \mathcal{L}_\infty$ [15].

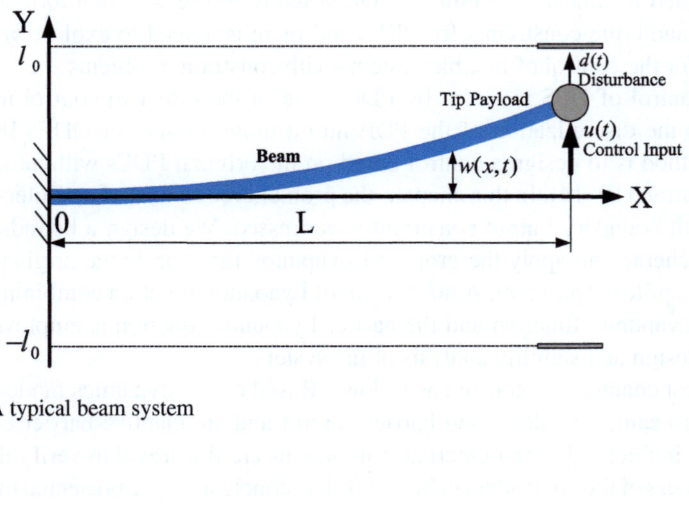

Fig. 4.1 A typical beam system

4.2.1 Model-Based Boundary Control

In this section, a novel barrier Lyapunov function is proposed and guarantees that the boundary output constraint is not violated, when the system parameters M, T and EI are all known. The model-based control is proposed as

$$u(t) = -\zeta(t)\left\{k_1 + \frac{Mw(L,t)\dot{w}(L,t)}{l_0^2 - [w(L,t)]^2}\right\} \Big/ \ln\frac{2l_0^2}{l_0^2 - [w(L,t)]^2}$$
$$-k_2\zeta(t) + M\left[c\dot{w}'''(L,t) - \dot{w}'(L,t)\right] + Tw'(L,t)$$
$$-EIw'''(L,t) - \mathrm{sgn}[\zeta(t)]\bar{d}, \tag{4.4}$$

where k_1 and k_2 are two positive control gains, c is a positive constant, and $\zeta(t)$ is defined as

$$\zeta(t) = \dot{w}(L,t) + w'(L,t) - cw'''(L,t). \tag{4.5}$$

Consider the following Lyapunov candidate function as

$$V(t) = V_1(t) + V_2(t) + V_3(t), \tag{4.6}$$

where the energy term $V_1(t)$, the barrier term $V_2(t)$, and the crossing term $V_3(t)$ are defined as

$$V_1(t) = \frac{a}{2}\rho\int_0^L [\dot{w}(x,t)]^2 dx + \frac{a}{2}EI\int_0^L [w''(x,t)]^2 dx$$
$$+\frac{a}{2}T\int_0^L [w'(x,t)]^2 dx, \tag{4.7}$$

$$V_2(t) = \frac{aM}{2}\zeta^2(t)\ln\frac{2l_0^2}{l_0^2 - w^2(L,t)}, \tag{4.8}$$

$$V_3(t) = b\rho\int_0^L x\dot{w}(x,t)w'(x,t)dx, \tag{4.9}$$

where a and b are positive constants, and $\ln(\bullet)$ denotes the natural logarithm of $(\bullet)$.

Lemma 4.1 *The Lyapunov candidate function given by (4.6) is upper and lower bounded as*

$$0 \le \lambda_1[V_1(t) + V_2(t)] \le V(t) \le \lambda_2[V_1(t) + V_2(t)], \tag{4.10}$$

where λ_1 and λ_2 are two positive constants defined as

$$\lambda_1 = 1 - \frac{2b\rho L}{\min(a\rho, aT)}, \tag{4.11}$$

$$\lambda_2 = 1 + \frac{2b\rho L}{\min(a\rho, aT)}, \tag{4.12}$$

provided

$$0 < b < \frac{\min(a\rho, aT)}{2\rho L}. \tag{4.13}$$

Proof Substituting (2.3) into (4.9) yields

$$|V_3(t)| \le b\rho L \int_0^L [\dot{w}(x,t)]^2 dx + b\rho L \int_0^L [w'(x,t)]^2 dx \le \alpha_1 V_1(t), \tag{4.14}$$

where

$$\alpha_1 = \frac{2b\rho L}{\min(a\rho, aT)}. \tag{4.15}$$

Then, we obtain

$$-\alpha_1 V_1(t) \le V_3(t) \le \alpha_1 V_1(t). \tag{4.16}$$

Considering b is a small positive weighting constant satisfying $0 < b < \min(a\rho, aT)/(2\rho L)$, we can obtain

$$\alpha_2 = 1 - \alpha_1 = 1 - \frac{2b\rho L}{\min(a\rho, aT)} > 0, \tag{4.17}$$

$$\alpha_3 = 1 + \alpha_1 = 1 + \frac{2b\rho L}{\min(a\rho, aT)} > 1. \tag{4.18}$$

Then,we further have

$$0 \le \alpha_2 V_1(t) \le V_1(t) + V_3(t) \le \alpha_3 V_1(t). \tag{4.19}$$

Given the Lyapunov function candidate (4.6), we obtain

$$0 \le \lambda_1[V_1(t) + V_2(t)] \le V_1(t) + V_2(t) + V_3(t) \le \lambda_2[V_1(t) + V_2(t)], \tag{4.20}$$

where $\lambda_1 = \min(\alpha_2, 1) = \alpha_2$ and $\lambda_2 = \max(\alpha_3, 1) = \alpha_3$ are positive constants.

Lemma 4.2 *If the control law (4.4) is utilized, the time derivative of the Lyapunov functional candidate (4.6) is upper bounded with*

$$\dot{V}(t) \leq -\lambda V(t), \tag{4.21}$$

where $\lambda > 0$.

Proof Differentiating (4.6) with respect to time leads to

$$\dot{V}(t) = \dot{V}_1(t) + \dot{V}_2(t) + \dot{V}_3(t). \tag{4.22}$$

Substituting the governing equation (4.1) and using the boundary conditions and Lemma 2.4, we have

$$\begin{aligned}
\dot{V}(t) \leq & -\frac{b\rho}{2}\int_0^L [\dot{w}(x,t)]^2 dx - \left(\frac{3bEI}{2} - \frac{|acT - bEIL|L}{\delta_3}\right) \\
& \times \int_0^L [w''(x,t)]^2 dx - \frac{bT}{2}\int_0^L [w'(x,t)]^2 dx \\
& - ak_2\zeta^2(t)\ln\frac{2l_0^2}{l_0^2 - [w(L,t)]^2} - a\left(k_1 - \frac{T}{2}\right)\zeta^2(t) \\
& - \left(\frac{ac^2T}{2} - \delta_3|acT - bEIL| - a\delta_2|cT - EI|\right) \\
& \times [w'''(L,t)]^2 - \frac{T(a - bL)}{2}[w'(L,t)]^2 \\
& - \left(\frac{aT}{2} - \frac{b\rho L}{2} - \frac{a|cT - EI|}{\delta_2}\right)[\dot{w}(L,t)]^2 \\
\leq & -\lambda_3\left[V_1(t) + V_2(t)\right],
\end{aligned} \tag{4.23}$$

where δ_2 and δ_3 are two positive constants, and λ_3 is defined as

$$\lambda_3 = \min\left(\frac{b\rho}{a\rho}, \frac{3b}{a} - \frac{2|acT - bEIL|L}{a\delta_3 EI}, \frac{bT}{aT}, \frac{2k_2}{M}\right) > 0, \tag{4.24}$$

The control parameters are chosen to satisfy following conditions

$$k_1 - \frac{T}{2} \geq 0, \tag{4.25}$$

$$\frac{ac^2T}{2} - \delta_3|acT - bEIL| - a\delta_2|cT - EI| \geq 0, \tag{4.26}$$

$$a - bL \geq 0, \tag{4.27}$$

$$\frac{aT}{2} - \frac{b\rho L}{2} - \frac{a|cT - EI|}{\delta_2} \geq 0. \tag{4.28}$$

Combining (4.10) and (4.23), we have

$$\dot{V}(t) \leq -\lambda V(t), \tag{4.29}$$

where $\lambda = \lambda_3/\lambda_2 > 0$.

Theorem 4.1 *For the system dynamics described by (4.1) and boundary conditions (4.2) to (4.3), using the proposed control (4.4), given that the initial conditions are bounded, we can conclude that:*

(i) the closed-loop system is exponentially stable;
(ii) if the initial boundary condition $w(L,0)$ belongs to the set $-l_0 < w(L,0) < l_0$, then $w(L,t)$ will remain in the open set $-l_0 < w(L,t) < l_0$, $\forall t \in [0,\infty)$.

Proof Multiplying (4.29) by $e^{\lambda t}$ yields

$$\frac{d}{dt}(V(t)e^{\lambda t}) \leq 0. \tag{4.30}$$

Integrating of the above inequality, we obtain

$$V(t) \leq V(0)e^{-\lambda t}. \tag{4.31}$$

which implies $V(t)$ is bounded. Utilizing (2.7), we have

$$\frac{a}{2L}Tw^2(x,t) \leq \frac{a}{2}T\int_0^L [w'(x,t)]^2 dx \leq V_1(t). \tag{4.32}$$

Appropriately rearranging the terms of the above inequality, we obtain $w(x,t)$ is exponentially stable as follows

$$|w(x,t)| \leq \sqrt{\frac{2L}{aT\lambda_1}(V(0))}, \quad \forall (x,t) \in [0,L] \times [0,\infty). \tag{4.33}$$

Furthermore, we have

$$\lim_{t\to\infty} |w(x,t)| = 0, \quad \forall x \in [0,L]. \tag{4.34}$$

From (4.34), it can be seen that $w(x,t)$ will converge exponentially to zero at the rate of convergence λ as $t \to \infty$. Therefore, there exists a time T_0, such that for all $t \geq T_0$, the boundary output converges to zero.

Remark 4.1 All the signals in our proposed control (4.4) and (4.35) can be measured by sensors or obtained by a backward difference algorithm. $w(L,t)$ can be sensed by a laser displacement sensor at the right boundary of the beam, $w'(L,t)$ can be measured by an inclinometer, $w'''(L,t)$ can be obtained by a shear force sensor, $\dot{w}(L,t)$ and $\dot{w}'''(L,t)$ can be calculated with a backward difference algorithm. It needs to be noted that differentiating twice and thrice the position $w(L,t)$ with respect to time to obtain $\ddot{w}(L,t)$ and $\dddot{w}(L,t)$ respectively, are undesirable in practice since the existing noise will be amplified. Therefore, in the control design, we just use the measurable signals and their first derivative as the feedback signal.

4.2.2 Adaptive Boundary Control

In this section, when the system parameters M, T and EI are all unknown, the adaptive barrier control is proposed as

$$\begin{aligned} u(t) = & -\zeta(t)\left\{k_1 + \frac{\widehat{M}(t)w(L,t)\dot{w}(L,t)}{l_0^2 - [w(L,t)]^2}\right\} \Big/ \ln\frac{2l_0^2}{l_0^2 - [w(L,t)]^2} \\ & -k_2\zeta(t) + \widehat{M}(t)\left[c\dot{w}'''(L,t) - \dot{w}'(L,t)\right] \\ & +\widehat{T}(t)w'(L,t) - \widehat{EI}(t)w'''(L,t) - \text{sgn}[\zeta(t)]\bar{d}, \end{aligned} \tag{4.35}$$

where k_1 and k_2 are two positive control gains, c is a positive constant, and $\zeta(t)$ is defined as in (4.5). $\widehat{M}(t)$, $\widehat{T}(t)$ $\widehat{EI}(t)$ are the estimates of M, T and EI. Let $\widetilde{M}(t) = M - \widehat{M}(t)$, $\widetilde{T}(t) = T - \widehat{T}(t)$ and $\widetilde{EI}(t) = EI - \widehat{EI}(t)$. The adaption laws are designed as

$$\begin{aligned} \dot{\widehat{M}}(t) = & a\gamma_1^{-1}\left[\dot{w}'(L,t) - c\dot{w}'''(L,t)\right]\zeta(t)\ln\frac{2l_0^2}{l_0^2 - w^2(L,t)} \\ & +a\gamma_1^{-1}\frac{w(L,t)\dot{w}(L,t)\zeta^2(t)}{l_0^2 - [w(L,t)]^2} - \sigma_1\gamma_1^{-1}\widehat{M}(t), \end{aligned} \tag{4.36}$$

$$\dot{\widehat{T}}(t) = -a\gamma_2^{-1}w'(L,t)\zeta(t)\ln\frac{2l_0^2}{l_0^2 - w^2(L,t)} - \sigma_2\gamma_2^{-1}\widehat{T}(t), \tag{4.37}$$

$$\dot{\widehat{EI}}(t) = a\gamma_3^{-1}w'''(L,t)\zeta(t)\ln\frac{2l_0^2}{l_0^2 - w^2(L,t)} - \sigma_3\gamma_3^{-1}\widehat{EI}(t), \tag{4.38}$$

where γ_1–γ_3 and σ_1–σ_3 are positive constants.

Remark 4.2 The σ modification term is introduced to improve the robustness of the closed-loop system [16, 17]. Without such a modification term, the estimates $\widehat{M}(t)$, $\widehat{T}(t)$, $\widehat{EI}(t)$ might drift to very large values, which will result in a variation of a high-gain control scheme [18–20].

Consider the following Lyapunov functional candidate

$$V_a(t) = V(t) + \frac{\gamma_1}{2}\widetilde{M}^2(t) + \frac{\gamma_2}{2}\widetilde{T}^2(t) + \frac{\gamma_3}{2}\widetilde{EI}^2(t), \tag{4.39}$$

where $V(t)$ is defined as in (4.6).

Lemma 4.3 *The Lyapunov candidate function given by (4.39) is upper and lower bounded as*

$$\begin{aligned} 0 \le \mu_1\left[V_1(t) + V_2(t) + \widetilde{M}^2(t) + \widetilde{T}^2(t) + \widetilde{EI}^2(t)\right] & \le V_a(t) \\ \le \mu_2\left[V_1(t) + V_2(t) + \widetilde{M}^2(t) + \widetilde{T}^2(t) + \widetilde{EI}^2(t)\right], & \end{aligned} \tag{4.40}$$

where μ_1 *and* μ_2 *are two positive constants.* $\mu_1 = \min\left(\lambda_1, \frac{\gamma_1}{2}, \frac{\gamma_2}{2}, \frac{\gamma_3}{2}\right)$ *and* $\mu_2 = \max\left(\lambda_2, \frac{\gamma_1}{2}, \frac{\gamma_2}{2}, \frac{\gamma_3}{2}\right)$.

Lemma 4.4 *If the control law (4.35) and the adaption laws (4.36)–(4.38) are utilized, the time derivative of the Lyapunov functional candidate (4.42) is upper bounded with*

$$\dot{V}_a(t) \leq -\mu V_a(t) + \varepsilon, \tag{4.41}$$

where $\mu > 0$ *and* $\varepsilon > 0$.

Proof $\dot{V}_a(t)$ can be obtained as

$$\begin{aligned}\dot{V}_a(t) = {} & \dot{V}(t) + \gamma_1 \widetilde{M}(t)\dot{\widetilde{M}}(t) + \gamma_2 \widetilde{T}(t)\dot{\widetilde{T}}(t) \\ & + \gamma_3 \widetilde{EI}(t)\dot{\widetilde{EI}}(t).\end{aligned} \tag{4.42}$$

Since the control law is only involved in $\dot{V}_2(t)$, then substituting the control law (4.35) into $\dot{V}_2(t)$, we have

$$\begin{aligned}\dot{V}_2(t) \leq {} & -ak_1\zeta^2(t) - ak_2\zeta^2(t)\ln\frac{2l_0^2}{l_0^2 - [w(L,t)]^2} \\ & + a\widetilde{M}(t)\zeta^2(t)\frac{w(L,t)\dot{w}(L,t)}{l_0^2 - [w(L,t)]^2} \\ & + a\widetilde{M}(t)\left[\dot{w}'(L,t) - c\dot{w}'''(L,t)\right]\zeta(t)\ln\frac{2l_0^2}{l_0^2 - w^2(L,t)} \\ & - a\widetilde{T}(t)w'(L,t)\zeta(t)\ln\frac{2l_0^2}{l_0^2 - w^2(L,t)} \\ & + a\widetilde{EI}(t)w'''(L,t)\zeta(t)\ln\frac{2l_0^2}{l_0^2 - w^2(L,t)}.\end{aligned} \tag{4.43}$$

Then substituting (4.43) into (4.42), and applying the adaption laws (4.36)–(4.38), we have

$$\begin{aligned}\dot{V}_a(t) & \leq -\lambda_3\left[V_1(t) + V_2(t)\right] + \sigma_1\widetilde{M}(t)\widehat{M}(t) + \sigma_2\widetilde{T}(t)\widehat{T}(t) + \sigma_3\widetilde{EI}(t)\widehat{EI}(t) \\ & \leq -\mu_3\left[V_1(t) + V_2(t) + \widetilde{M}^2(t) + \widetilde{T}^2(t) + \widetilde{EI}^2(t)\right] + \varepsilon,\end{aligned} \tag{4.44}$$

where

$$\mu_3 = \min\left(\lambda_3, \frac{\sigma_1}{2}, \frac{\sigma_2}{2}, \frac{\sigma_3}{2}\right) > 0, \tag{4.45}$$

$$\varepsilon = \frac{\sigma_1}{2}M^2 + \frac{\sigma_2}{2}T^2 + \frac{\sigma_3}{2}EI^2. \tag{4.46}$$

Combining (4.40) and (4.48), we have

$$\dot{V}_a(t) \leq -\mu V_a(t) + \varepsilon, \tag{4.47}$$

where $\mu = \mu_3/\mu_2 > 0$.

Theorem 4.2 *For the system dynamics described by (4.1) and boundary conditions (4.2) to (4.3), when the system parameters are unknown, using the proposed control (4.35) and the adaption laws (4.36)–(4.38), given that the initial conditions are bounded, we can conclude that:*

(i) the closed-loop system is uniform ultimate bounded;
(ii) all closed-loop signals are bounded;
(iii) if the initial boundary condition $w(L,0)$ belongs to the set $-l_0 < w(L,0) < l_0$, then $w(L,t)$ will remain in the open set $-l_0 < w(L,t) < l_0$, $\forall t \in [0,\infty)$.

Proof Applying the similar procedures as the proof for Theorem 4.1, we have

$$\frac{a}{2L}Tw^2(x,t) \leq \frac{a}{2}T\int_0^L [w'(x,t)]^2 dx \leq V_1(t) \leq \frac{1}{\lambda_1}V_a(t). \tag{4.48}$$

Appropriately rearranging the terms of the above inequality, we obtain $w(x,t)$ is uniformly bounded as follows

$$|w(x,t)| \leq \sqrt{\frac{2L}{aT\lambda_1}\left(V_a(0) + \frac{\varepsilon}{\mu}\right)}, \tag{4.49}$$

$$\frac{\gamma_1}{2}\widetilde{M}^2(t) \leq V_a(t) \leq V_a(0)e^{-\mu t} + \frac{\varepsilon}{\mu}, \tag{4.50}$$

$$|\widetilde{M}(t)| \leq \sqrt{\frac{2}{\gamma_1}[V_a(0)e^{-\mu t} + \frac{\varepsilon}{\mu}]}. \tag{4.51}$$

In the similar method, we obtain $|\widetilde{T}(t)| \leq \sqrt{\frac{2}{\gamma_2}[V_a(0)e^{-\mu t} + \frac{\varepsilon}{\mu}]}$ and $|\widetilde{EI}(t)| \leq \sqrt{\frac{2}{\gamma_3}[V_a(0)e^{-\mu t} + \frac{\varepsilon}{\mu}]}$ $\forall (x,t) \in [0,L] \times [0,\infty)$.

Furthermore, we have $\lim_{t\to\infty}|w(x,t)| = \sqrt{\frac{2L\varepsilon}{aT\lambda_1\mu}}$, $\lim_{t\to\infty}|\widetilde{M}(t)| = \sqrt{\frac{2\varepsilon}{\gamma_1\mu}}$, $\lim_{t\to\infty}|\widetilde{T}(t)| = \sqrt{\frac{2\varepsilon}{\gamma_2\mu}}$ and $\lim_{t\to\infty}|\widetilde{EI}(t)| = \sqrt{\frac{2\varepsilon}{\gamma_3\mu}}$, $\forall x \in [0,L]$.

From $\lim_{t\to\infty}|w(x,t)| = \sqrt{\frac{2L\varepsilon}{aT\lambda_1\mu}}$, it can be seen that $w(x,t)$ will converge exponentially to a small neighborhood of zero. Therefore, there exists a time T_0, such that for all $t \geq T_0$, the boundary output converges to a small neighborhood around zero. Using the method of proof in Theorem 4.1, we can also conclude the boundary control (4.35) is also bounded $\forall t \in [0,\infty)$.

From (4.48) and (4.40), we have $V_1(t) + V_2(t) + \widetilde{M}^2(t) + \widetilde{T}^2(t) + \widetilde{EI}^2(t)$ are positive and bounded $\forall t \in [0,\infty)$, and $V_a(t)$ is also bounded $\forall t \in [0,\infty)$. Since $V_a(t)$, $V_1(t)$ and $V_2(t)$ are positive functions, we know that $V_2(t)$ is also bounded $\forall t \in$

Table 4.1 Parameters of the flexible beam

Parameter	Description	Value
L	Length of beam	1 m
EI	Bending stiffness	5 Nm^2
T	Tension	6 N
ρ	Mass per unit length	1 kg/m
M	Mass of the tip payload	5 kg
l_0	Boundary output constraint	0.05 m

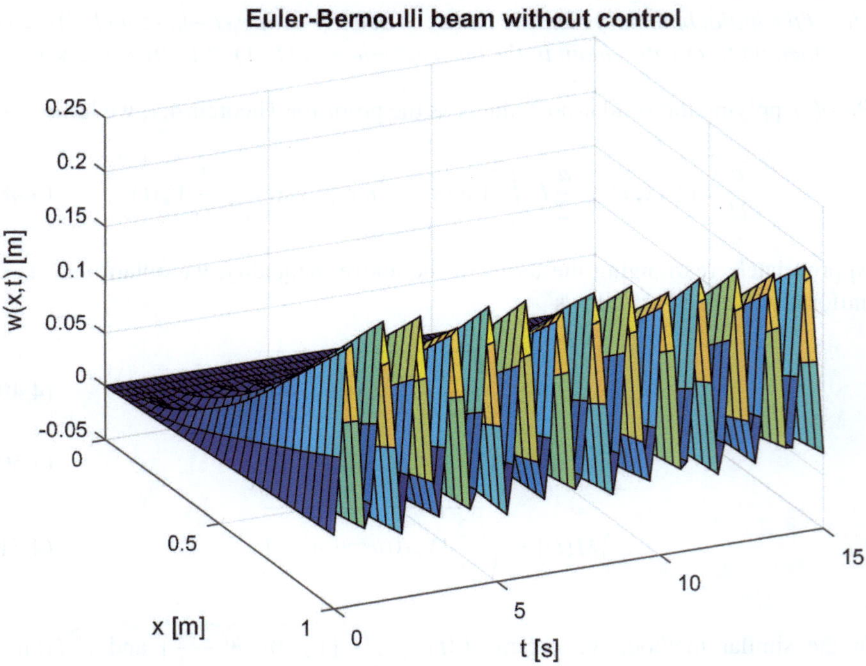

Fig. 4.2 Displacement of the beam without control

$[0, \infty)$. From the definition of $V_2(t)$, we know that $V_2(t) \to \infty, as \mid w(L, t) \mid \to l_0$. Consequently, we know that $|w(L, t)| \neq l_0$. Given that $-l_0 < w(L, 0) < l_0$, we infer that $w(L, t)$ remains in the set $-l_0 < w(L, t) < l_0$, $\forall t \in [0, \infty)$.

4.3 Numerical Simulations

The effectiveness of the proposed adaptive barrier control is illustrated by numerical simulations by using the finite difference method. The corresponding initial conditions of the Euler-Bernoulli beam are given as $w(x, 0) = \frac{x}{55}$, $\dot{w}(x, 0) = 0, x \in [0, L]$.

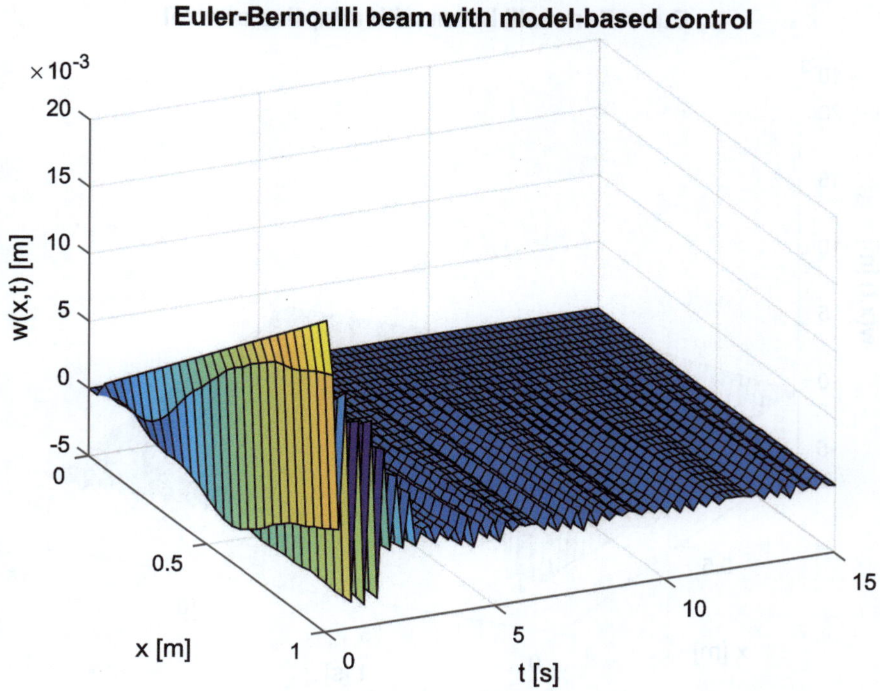

Fig. 4.3 Displacement of the beam with model-based barrier control (4.4)

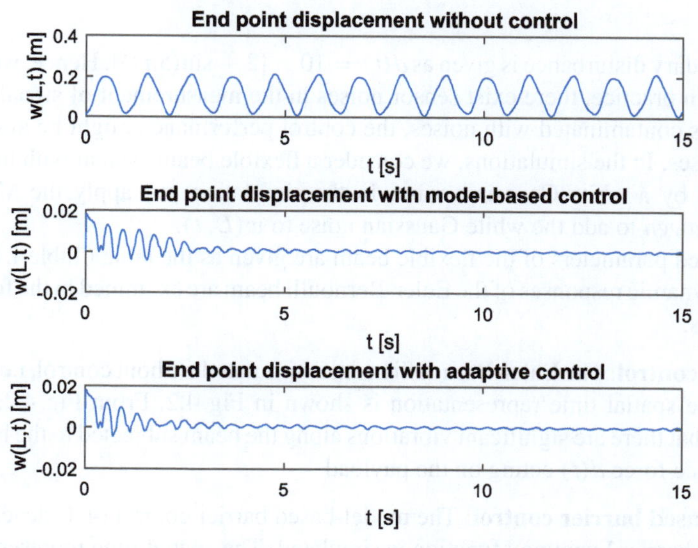

Fig. 4.4 Boundary outputs of the beam: without control, with model-based barrier control (4.4), with adaptive barrier control (4.35)

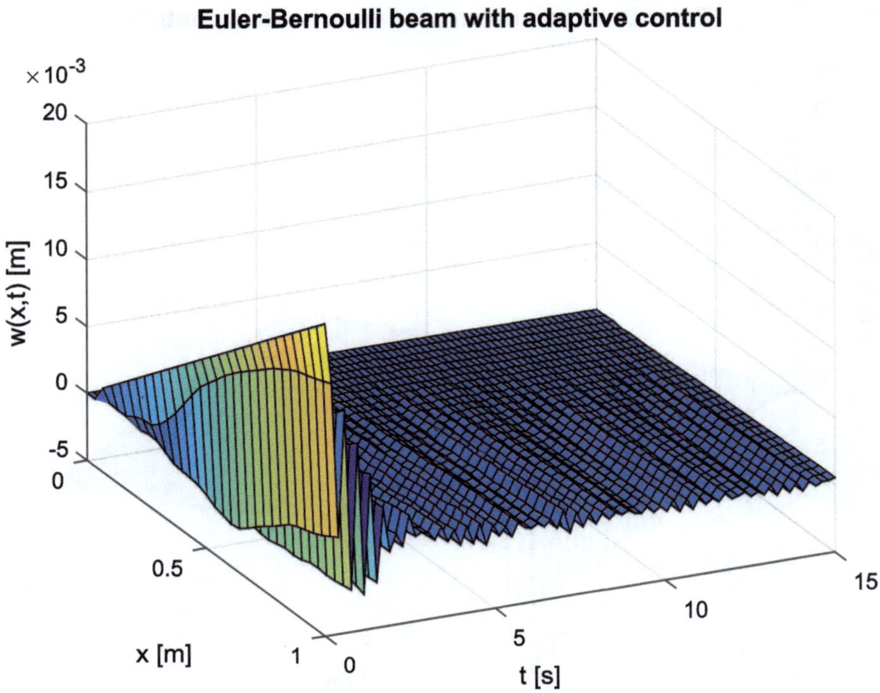

Fig. 4.5 Displacement of the beam with adaptive barrier control (4.35)

The boundary disturbance is given as $d(t) = 10 \times [2 + \sin(5\pi t)]$. Hence, we choose $\bar{d} = 30$. In practice, there exist sensor noises in the measurement of signals. When $w(L, t)$ is contaminated with noises, the control performance might be affected by those noises. In the simulations, we consider a flexible beam system with the noises produced by a white Gaussian noise. In the simulation, we apply the MATLAB function *awgn* to add the white Gaussian noise to $w(L, t)$.

Detailed parameters of the flexible beam are given as the table (Table 4.1).

The dynamic responses of the Euler-Bernoulli beam are examined in the following four cases.

Without control: The Euler-Bernoulli beam is simulated without control, i.e., $u(t) = 0$, and the spatial time representation is shown in Fig. 4.2. From Fig. 4.2, we can observe that there are significant vibrations along the beam subjected to the boundary disturbance force $d(t)$ acting on the payload.

Model-based barrier control: The model-based barrier control (4.4) developed by using the barrier Lyapunov function is simulated. The spatial time representation is shown in Fig. 4.3, and the boundary output (tip payload) is shown in Fig. 4.4. It can be seen that the designed control is able to suppress the vibration of the beam exponentially within 8 seconds, when the system is subjected to the external disturbance $d(t)$. In addition, as shown in Fig. 4.4, the model-based barrier control (4.4) guarantees

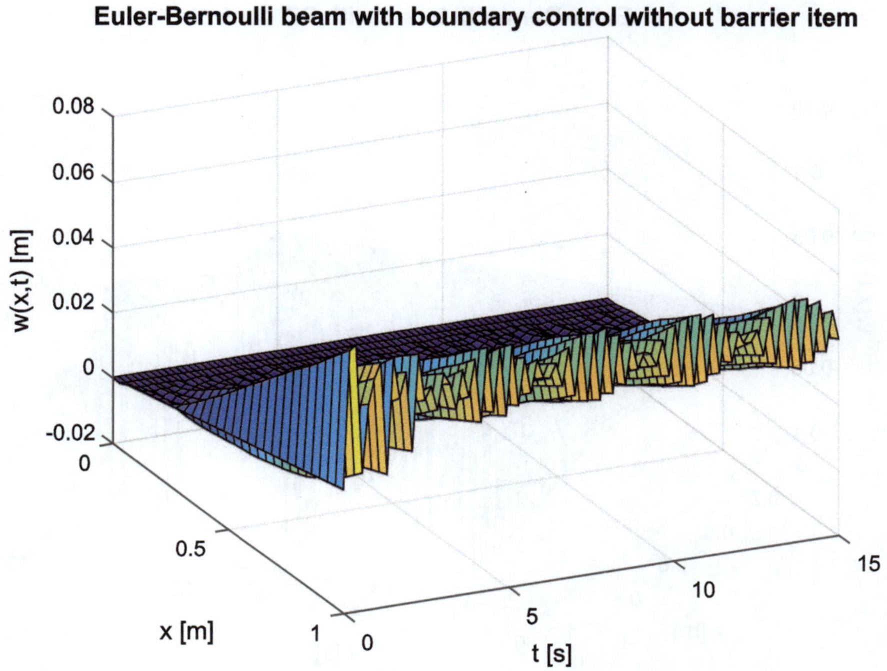

Fig. 4.6 Displacement of the beam with boundary control without the barrier item

that the boundary output satisfies $| w(L,t) | \leq l_0$. From the simulation results, we also can conclude the proposed control still can work well under the effects of the noise.

Adaptive barrier control: The adaptive barrier control (4.35) and adaption laws (4.36)–(4.38) is simulated. The simulation results are shown in Figs. 4.4 and 4.5.

Boundary control without the barrier term: For comparison, displacements of the beam with the following boundary control proposed in [15],

$$u(t) = -ku_a(t) - EIw'''(L,t) + Tw'(L,t) \\ -\hat{d}(t) - M[\dot{w}'(L,t) - \dot{w}'''(L,t)], \tag{4.52}$$

is simulated in Fig. 4.6, where $u_a(t)$ is defined as $u_a(t) = \dot{w}(L,t) - w'''(L,t) + w'(L,t)$. $\hat{d}(t)$ is the estimate of the upper bound of the boundary disturbance $\bar{d}$, and the disturbance observer $\dot{\hat{d}}(t) = \gamma u_a(t) - \xi_d \gamma \hat{d}(t)$, where γ and ξ_d are two positive constants.

Comparing with the proposed adaptive barrier control (4.4), the barrier term $\{\bullet\}/\ln \frac{2l_0^2}{l_0^2 - [w(L,t)]^2}$ is removed in (4.52). Figure 4.6 illustrates that the boundary control (4.52) is able to stabilize the beam at the small neighborhood of its equilibrium position. However, as shown in Fig. 4.3, the adaptive barrier control (4.4) guarantees

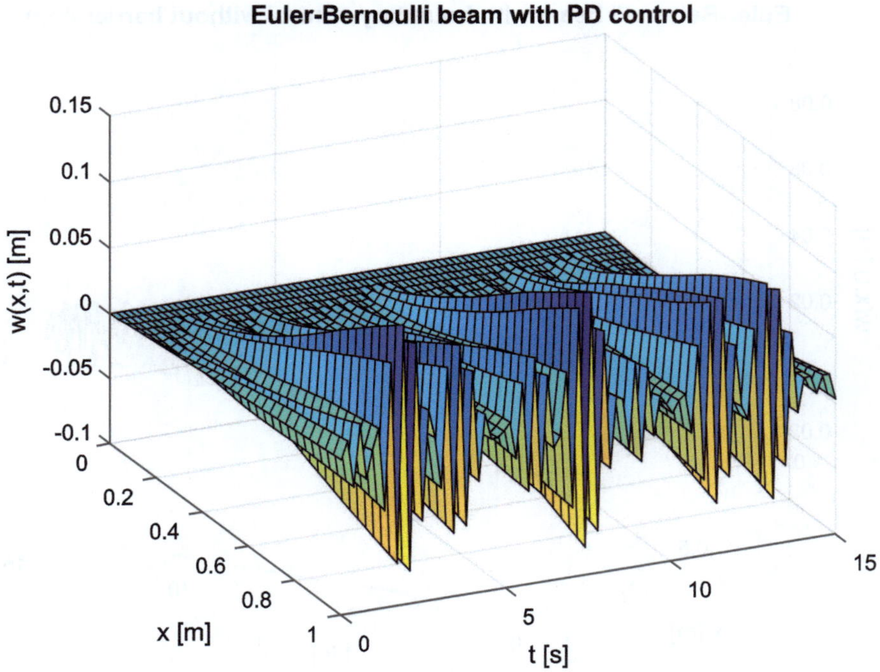

Fig. 4.7 Displacement of the beam with PD control

that the boundary output position can be converged to $|w(L, t)| \leq l_0$, while boundary control (4.52) cannot guarantee $|w(L, t)| \leq l_0$.

PD control: For comparison, displacements of the beam with PD control $u(t) = -k_p w(L, t) - k_d \dot{w}(L, t)$ is presented in Fig. 4.7, where k_p is chosen as 0.9×10^3 and k_d is chosen as 0.1. In the Fig. 4.7, PD control laws can suppress the vibrations and regulate the system at the small neighborhood of its equilibrium position. However, it can be found that the PD control cannot ensure the beam's end point deflection $|w(L, t)| < l_0$.

In conclusion, the simulation results displayed in Figs. 4.3–4.5 show that both the model-based barrier control (4.4) and adaptive barrier control (4.35) are effective in handling the boundary output constraint, as well as regulating the vibrations of the beam system, illustrating that the proposed control can improve the performance of the beam system.

4.4 Conclusion

In this chapter, both model-based control and adaptive barrier control have been developed for an Euler-Bernoulli beam system with boundary output constraint. The vibration of the beam has been suppressed without violation of the constraint.

By employing a novel barrier Lyapunov function, the exponential stability of the closed-loop beam system has been proved when the system model is exactly known. When there are parameter uncertainties, boundedness of the closed-loop beam system has been achieved via the adaptive boundary control. The simulation results have shown that the proposed control can suppress the vibration of the beam with a good performance.

References

1. Q.C. Nguyen, K.-S. Hong, Simultaneous control of longitudinal and transverse vibrations of an axially moving string with velocity tracking. J. Sound Vib. **331**(13), 3006–3019 (2012)
2. W. He, S. Zhang, S.S. Ge, Boundary control of a flexible riser with application to marine installation. IEEE Trans. Ind. Electron. **60**(12), 5802–5810 (2013)
3. L. Lu, Z. Chen, B. Yao, Q. Wang, A two-loop performance oriented tip tracking control of a linear motor driven flexible beam system with experiments. IEEE Trans. Ind. Electron. **60**(3), 1011–1022 (2013)
4. W. He, S.S. Ge, S. Zhang, Adaptive boundary control of a flexible marine installation system. Automatica **47**(12), 2728–2734 (2011)
5. K.P. Tee, S.S. Ge, E. Tay, Barrier Lyapunov functions for the control of output-constrained nonlinear systems. Automatica **45**(4), 918–927 (2009)
6. B. Ren, S.S. Ge, K.P. Tee, T. Lee, Adaptive neural control for output feedback nonlinear systems using a barrier Lyapunov function. IEEE Trans. Neural Netw. **21**(8), 1339–1345 (2010)
7. K.P. Tee, B. Ren, S.S. Ge, Control of nonlinear systems with time-varying output constraints. Automatica **47**(11), 2511–2516 (2011)
8. H.-N. Wu, J.-W. Wang, H.-X. Li, Design of distributed H_∞ fuzzy controllers with constraint for nonlinear hyperbolic PDE systems. Automatica **48**(10), 2535–2543 (2012)
9. H.-N. Wu, H.-X. Li, Finite-dimensional constrained fuzzy control for a class of nonlinear distributed process systems. IEEE Trans. Syst. Man Cybern. Part B (Cybern.) **37**(5), 1422–1430 (2007)
10. B. Ren, J.-M. Wang, M. Krstic, Stabilization of an ODE-Schrödinger cascade. Syst. Control Lett. **62**(6), 503–510 (2013)
11. J.-M. Wang, B. Ren, M. Krstic, Stabilization and gevrey regularity of a Schrödinger equation in boundary feedback with a heat equation. IEEE Trans. Autom. Control **57**(1), 179–185 (2012)
12. Q.C. Nguyen, K.-S. Hong, Transverse vibration control of axially moving membranes by regulation of axial velocity. IEEE Trans. Control Syst. Technol. **20**(4), 1124–1131 (2012)
13. K.-S. Hong, J. Bentsman, Direct adaptive control of parabolic systems: algorithm synthesis and convergence and stability analysis. IEEE Trans. Autom. Control **39**(10), 2018–2033 (1994)
14. W. He, S. Zhang, S.S. Ge, Adaptive boundary control of a nonlinear flexible string system. IEEE Trans. Control Syst. Technol. **22**(3), 1088–1093 (2014)
15. S.S. Ge, S. Zhang, W. He, Vibration control of an Euler-Bernoulli beam under unknown spatiotemporally varying disturbance. Int. J. Control **84**(5), 947–960 (2011)
16. Z. Li, C.-Y. Su, Neural-adaptive control of single-master-multiple-slaves teleoperation for coordinated multiple mobile manipulators with time-varying communication delays and input uncertainties. IEEE Trans. Neural Netw. Learn. Syst. **24**(9), 1400–1413 (2013)
17. P. Ioannou, J. Sun, *Robust Adaptive Control* (Prentice-Hall, New Jersey, 1996)
18. Z. Li, J. Li, Y. Kang, Adaptive robust coordinated control of multiple mobile manipulators interacting with rigid environments. Automatica **46**(12), 2028–2034 (2010)
19. C. Yang, Z. Li, J. Li, Trajectory planning and optimized adaptive control for a class of wheeled inverted pendulum vehicle models. IEEE Trans. Cybern. **43**(1), 24–36 (2013)
20. S.-L. Dai, C. Wang, F. Luo, Identification and learning control of ocean surface ship using neural networks. IEEE Trans. Ind. Inform. **8**(34), 801–810 (2012)

Chapter 5
Vibration Control of a Flexible Beam with Input Saturation

5.1 Introduction

In the previous works, the constraint of the input actuator is ignored in the control design. However, in the actual control engineering, the completely linear system does not exist, and almost all the systems have the input constraints due to the limitation of the actuators or the inherent physical constraints of the systems. In the implementation of the flexible beam system with active control, the limitations modeled as nonlinear saturation characteristics exist in the control input, such as the finite voltage which hydraulic motors can achieve, the finite capacity of the pump, and so on [1]. The saturation characteristic can make the system unstable and degrade the performance of the system. Therefore, the effect of the input saturation characteristic should be taken into consideration for control design.

In this chapter, the largest challenge is how to compensate for the nonlinear input saturation characteristic and to design an effective active control law for the flexible beam system. There are some methods to deal with the system with the input saturation, namely, low-gain control [2], anti-windup compensator [3], neural network control [4], model predictive control [5] and adaptive control [6, 7]. In [3], an anti-windup compensator is designed to deal with a linear system related to input saturation and to analyze the $\mathfrak{L}_2$ performance properties and the quadratic stability of this system. In [4], the authors propose a robust adaptive neural network control for a multiple-input-multiple-output (MIMO) unknown system with nonlinear input saturation and dead-zone using the variable structure control, backstepping and Lyapunov's direct method. In [8], a robust adaptive control is designed for nonlinear systems with input saturation, unknown parameters and unknown external disturbances by using the backstepping method. However, all the above mentioned works investigate the control problems for the ODE systems with input saturation. In this chapter, we are going to propose an auxiliary system in the control design of a beam to reduce the influence induced by the input saturation.

In the design of the control, the signals used to make up the control law can be measured by the sensors directly or obtained with the backward difference algorithm.

W. He and J. Liu, *Active Vibration Control and Stability Analysis of Flexible Beam Systems*, https://doi.org/10.1007/978-981-10-7539-1_5

However, the measured noise may be magnified with the difference algorithm. It is a crucial issue to be considered in the design procedure of the control. In [9], control laws are designed to suppress the vibration of the first two modes for a smart flexible cantilever beam using the periodic output feedback control technique. In [10], by estimating the states of the system, the output feedback frequency shaped sliding mode control is proposed to reduce the vibration of intelligent structure effectively. However, these works only study the systems which are modeled by ordinary differential equations. In this chapter, the problem related to the unmeasurable states is analyzed and discussed using output feedback boundary control based on the infinite dimensional dynamical model. The high-gain observers are designed to estimate the unmeasurable system states.

The rest chapter is organized as follows. The dynamics model of Euler-Bernoulli beam with input saturation is derived in Sect. 5.2. A boundary control is designed in Sect. 5.3. The numerical simulations are provided to verify the control effectiveness of the controllers in Sect. 5.4. The conclusions are summarized in Sect. 5.5.

5.2 Problem Formulation

As shown in Fig. 5.1, an actuator is designed at the top boundary of the flexible beam. The control implemented by the actuator is under the limitation of saturation.

In this chapter, we consider the flexible beam with input saturation which is common in practice, and the input saturation function is given as follows [1],

$$u(t) = \begin{cases} \mathrm{sgn}(u_0(t))u_{\max}, & |u_0(t)| \geq u_{\max} \\ u_0(t), & |u_0(t)| < u_{\max} \end{cases} \tag{5.1}$$

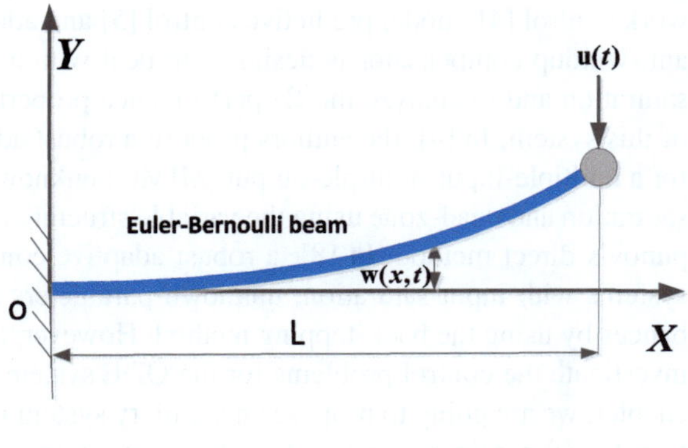

Fig. 5.1 A typical beam system

where u is the control input, u_0 is the designed control command, u_{max} is the saturation limit.

5.3 Control Design

The flexible beam system is described with the governing equation,

$$EIw''''(x,t) + \rho\ddot{w}(x,t) - Tw''(x,t) = 0 \tag{5.2}$$

and boundary conditions,

$$\begin{aligned} &w(0,t) = w'(0,t) = w''(L,t) = 0; \\ &-EIw'''(L,t) + Tw'(L,t) = u(t) - M_s\ddot{w}(L,t); \quad \forall t \in [0,\infty). \end{aligned} \tag{5.3}$$

Boundary control is designed to suppress the vibration of the flexible beam given by (5.2)–(5.3), and make the displacement of the beam converge to a small neighborhood of its original position. In this section, the boundary control law is designed with Lyapunov's direct method.

In order to suppress the vibration of the flexible beam, we propose the boundary control law as follows

$$u(t) = -\tanh[\dot{w}(L,t)] \tag{5.4}$$

Choose the Lyapunov candidate function as

$$V(t) = V_a(t) + V_b(t) \tag{5.5}$$

where

$$\begin{aligned} V_a(t) &= \frac{\rho}{2}\int_0^L [\dot{w}(x,t)]^2 dx + \frac{T}{2}\int_0^L [w'(x,t)]^2 dx \\ &\quad + \frac{EI}{2}\int_0^L [w''(x,t)]^2 dx \end{aligned} \tag{5.6}$$

$$V_b(t) = \frac{M_s}{2}\dot{w}^2(L,t) \tag{5.7}$$

Lemma 5.1 *The Lyapunov function represented by* (5.5)*, has lower bound as*

$$0 \le V(t) \tag{5.8}$$

Lemma 5.2 *The time derivative of the Lyapunov candidate function* (5.5) *is bounded with*

$$\dot{V}(t) \leq 0 \tag{5.9}$$

Proof Differentiating (5.5) with respect to time leads to

$$\dot{V}(t) = \dot{V}_a(t) + \dot{V}_b(t) \tag{5.10}$$

Using the governing equation (5.2) and integrating the first term of (5.10) by part, we obtain

$$\dot{V}_a(t) = -EIw'''(L,t)\dot{w}(L,t) + Tw'(L,t)\dot{w}(L,t) \tag{5.11}$$

Differentiating (5.7), we have

$$\dot{V}_b(t) = M_s\dot{w}(L,t)\ddot{w}(L,t) \tag{5.12}$$

Substituting (5.11), (5.12) into (5.10), we obtain

$$\begin{aligned} \dot{V}(t) &= \dot{w}(L,t)u(t) \\ &= -\dot{w}(L,t)\tanh[\dot{w}(L,t)] \\ &\leq 0 \end{aligned} \tag{5.13}$$

Therefore, by LaSalle's invariance principle [11], the boundary control can guarantee the uniform ultimate boundedness of the closed-loop system.

5.4 Numerical Simulations

In order to illustrate the effectiveness of the proposed boundary control law for the flexible beam, we present the numerical simulation. The parameters of the system are listed in Table 5.1.

Table 5.1 Parameters of the flexible beam system

Parameter	Description	Value	Unit
L	Length of the beam	1	m
M_s	Mass of the beam	5	kg
EI	Bending stiffness of the beam	5	Nm^2
ρ	Mass per unit length of the beam	1	kg/m
T	Tension of the beam	6	N

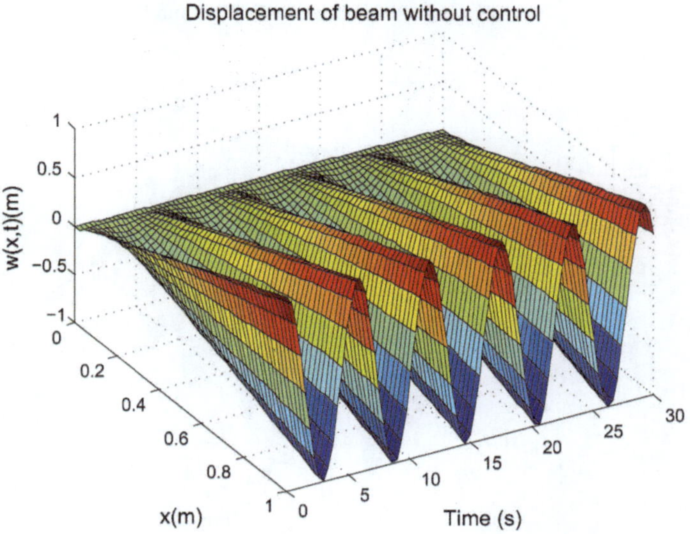

Fig. 5.2 Displacement of the beam without control

To demonstrate the effectiveness and practicability of the proposed control for a classical flexible beam system with input saturation, we choose the finite difference method to carry out the numerical simulation. For analyzing the effect of the proposed boundary control conveniently, we consider of three different cases as follows

(i) **Without control**: The flexible beam system is vibrating without control force, namely, in the boundary control of the dynamical model, $u(t) = 0$. In Fig. 5.2, we can see that the deflection of the flexible beam system is varying periodically with the amplitude which may cause serious damage to this system.

(ii) **PD control**: In Fig. 5.3, the simulation result illustrates the effect acted in the flexible beam system by the PD control $u_0(t) = 10w(L,t) + 1.5\dot{w}(L,t)$. The vibration of the beam system is reduced. However, the settling time of the closed-loop system is longer. The designed PD control command is shown in Fig. 5.4 and the PD control input is shown in Fig. 5.5.

(iii) **Boundary control**: The deflection of the flexible beam system with state boundary control is shown in Fig. 5.6. The proposed boundary control has reduced the vibration of the system and the displacement of the end point of the beam converge to a small neighborhood of zero. In addition, in Fig. 5.8, the proposed boundary control law is presented, the designed control command of the system is shown in Fig. 5.7.

According to the presented simulation results, we conclude that the proposed boundary control can reduce the vibration of the beam system effectively.

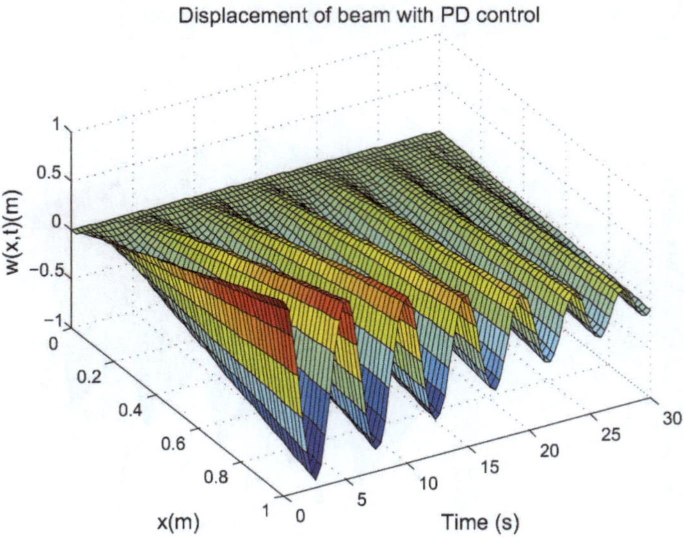

Fig. 5.3 Displacement of the beam system with PD control

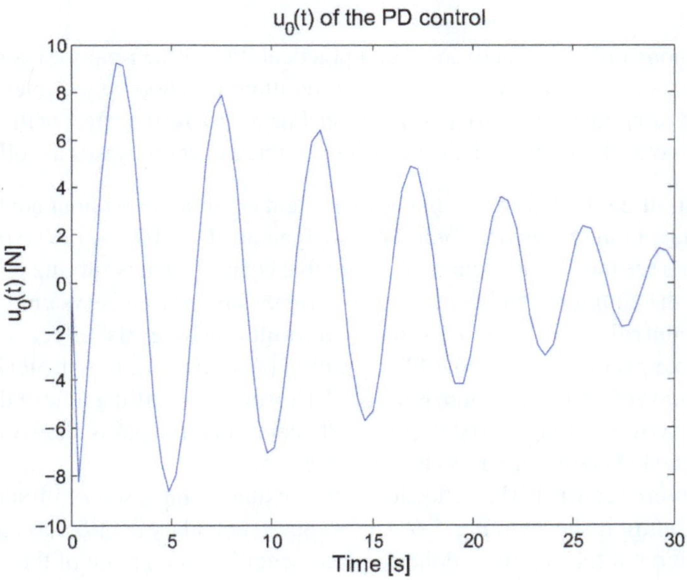

Fig. 5.4 Designed control command of the system with PD control

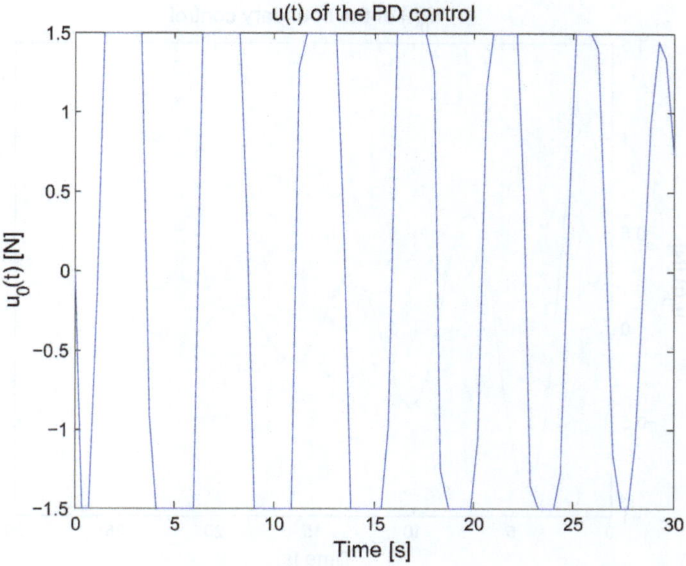

Fig. 5.5 The control input of the system with PD control

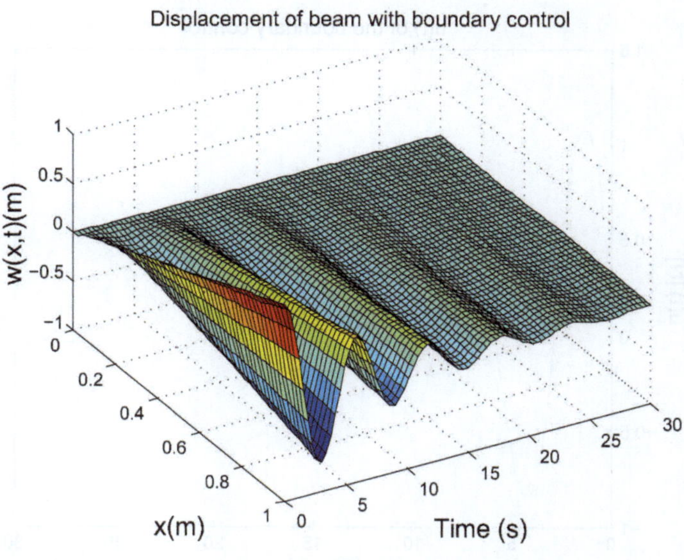

Fig. 5.6 Displacement of the beam system with boundary control

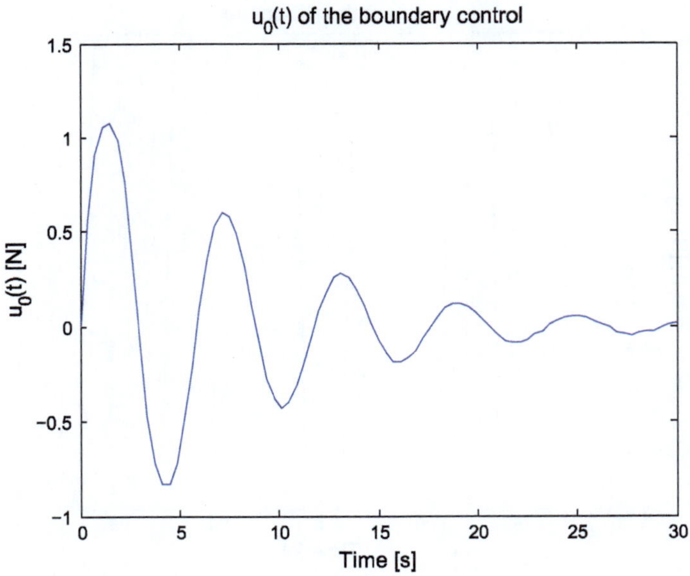

Fig. 5.7 Designed control command of the system with boundary control

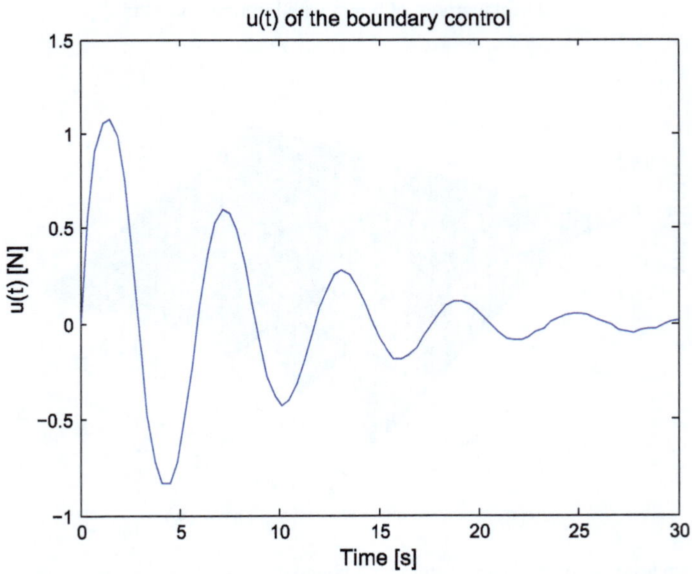

Fig. 5.8 The control input of the system with boundary control

5.5 Conclusion

This chapter investigates the vibration control problem of a flexible beam with input saturation. In addition, the state feedback boundary control is designed at the top of flexible beam via constructing a Lyapunov candidate function to analyze the convergence of the solution of closed-loop system. It is proved that the state of the closed-loop system converges to a small neighborhood of zero. Numerical simulations are provided to express the effectiveness of the boundary control law. The proposed control is implementable since all the signals in the control inputs can be measured directly by sensors or be obtained by a backward difference algorithm.

References

1. J. Zhou, C. Wen, *Adaptive backstepping control of uncertain systems: Nonsmooth nonlinearities, interactions or time-variations* (Springer, Berlin, Germany, 2008)
2. Z. Lin and A. Saberi, Low-and-high gain design technique for linear systems subject to input saturation-a direct method, in *Proceedings of the 35th IEEE Conference on Decision and Control*, vol. 4 (IEEE, 1996), pp. 4788–4793
3. G. Grimm, J. Hatfield, I. Postlethwaite, A.R. Teel, M.C. Turner, L. Zaccarian, Antiwindup for stable linear systems with input saturation: an lmi-based synthesis. IEEE Trans. Autom. Control **48**(9), 1509–1525 (2003)
4. M. Chen, S.S. Ge, B. Ren, Adaptive tracking control of uncertain mimo nonlinear systems with input constraints. Automatica **47**(3), 452–465 (2011)
5. V. Adetola, D. DeHaan, M. Guay, Adaptive model predictive control for constrained nonlinear systems. Syst. Control Lett. **58**(5), 320–326 (2009)
6. S. Karason, A. Annaswamy, Adaptive control in the presence of input constraints. IEEE Trans. Autom. Control **39**(11), 2325–2330 (1994)
7. A. Annaswamy, S. Karason, Discrete-time adaptive control in the presence of input constraints. Automatica **31**(10), 1421–1431 (1995)
8. C. Wen, J. Zhou, Z. Liu, H. Su, Robust adaptive control of uncertain nonlinear systems in the presence of input saturation and external disturbance. IEEE Trans. Autom. Control **56**(7), 1672–1678 (2011)
9. T. Manjunath, B. Bandyopadhyay, Vibration control of a smart structure using periodic output feedback technique. Asian J. Control **6**(1), 74–87 (2004)
10. A.J. Mehta, B. Bandyopadhyay, Frequency-shaped sliding mode control using output sampled measurements. IEEE Trans. Ind. Electron. **56**(1), 28–35 (2009)
11. O. Morgul, Stabilization and disturbance rejection for the beam equation. IEEE Trans. Autom. Control **46**(12), 1913–1918 (2001)

Chapter 6
Vibration Control of a Flexible Beam with Input Dead-Zone

6.1 Introduction

Control of nonlinear dynamical systems preceded by some non-smooth nonlinearities such as dead-zone, saturation, and backlash in the actuator has received much attention among the research area due to the diverse applications of mechanical systems in real-life. These non-smooth nonlinearities in actuators may degrade the performance of control system in practical control systems, and should be handled in the control design.

Dead-zone is one of the most important non-smooth nonlinearities arisen in servo valves and DC servo motors [1]. In fact, dead-zone nonlinearity is often unknown and time-variant. A number of researches focus on the control system represented by ordinary differential equations (ODEs) with input dead-zone nonlinearity. In [2], an adaptive control law is proposed for continuous-time linear dynamic systems with an unknown dead-zone in state space form. Later, a robust adaptive control scheme is designed in [1] for a class of special nonlinear systems with an unknown symmetrical dead-zone without constructing the dead-zone inverse. In [3], robust adaptive neural network (NN) control is proposed for a class of uncertain multiple-input-multiple-output (MIMO) nonlinear systems with both nonsymmetric input saturation and dead-zone. However, studies for control of flexible systems which are governed by partial differential equations (PDEs) with input dead-zone nonlinearity are still limited. In this chapter, we focus on the boundary control of PDE-based flexible beam systems with input dead-zone by using the Lyapunov theory.

The rest parts of this chapter are summarized as follows. The problem formulation and preliminaries of the chapter is presented for the flexible mechanical systems with the input dead-zone constraint in Sect. 6.2. In Sect. 6.3, we design a control law for analyzing the system, and use the Lyapunov's direct method to obtain the stability of the flexible beam system. The effectiveness of the proposed control approach is proved by the performance of the numerical simulation with finite difference method in Sect. 6.4. Section 6.5 is the conclusion of the chapter.

W. He and J. Liu, *Active Vibration Control and Stability Analysis of Flexible Beam Systems*, https://doi.org/10.1007/978-981-10-7539-1_6

6.2 Problem Formulation

In this chapter, we consider flexible beam systems with the input dead-zone constraint (Fig. 6.1).

The dead-zone nonlinearity [4] is shown in Fig. 6.2 and descried as follows

$$D(v) = \begin{cases} g_r(v), & v \geq b_r \\ 0, & b_l < v < b_r \\ g_l(v), & v \leq b_l \end{cases} \tag{6.1}$$

where $D(v)$ is the actual control input, $v(t)$ is the designed control command, b_r and b_l are the unknown bounded constants, $g_r(v)$ and $g_l(v)$ are the unknown smooth functions. The authors in [4] also give several assumptions as follows

Assumption 6.1 ([4]) The dead-zone output $D(v)$ is not available.

Assumption 6.2 ([4]) The dead-zone parameters, b_r and b_l, are unknown bounded constants, but their signs are known, i.e., $b_r > 0$ and $b_l < 0$.

Assumption 6.3 ([4]) The functions, $g_r(v)$ and $g_l(v)$, are smooth, and there exist unknown positive constants, k_{r0}, k_{r1}, k_{l0} and k_{l1}, such that

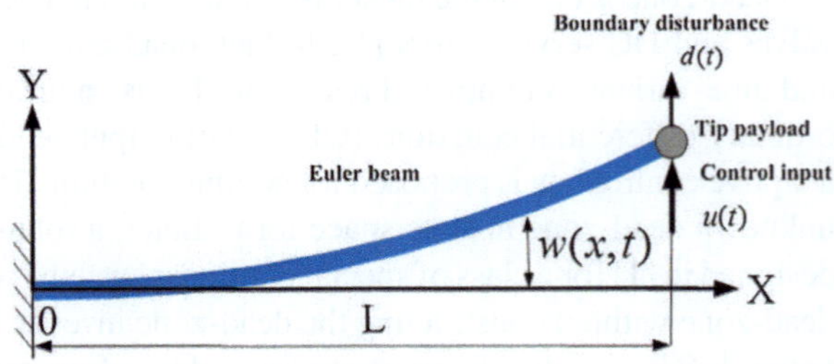

Fig. 6.1 The beam model

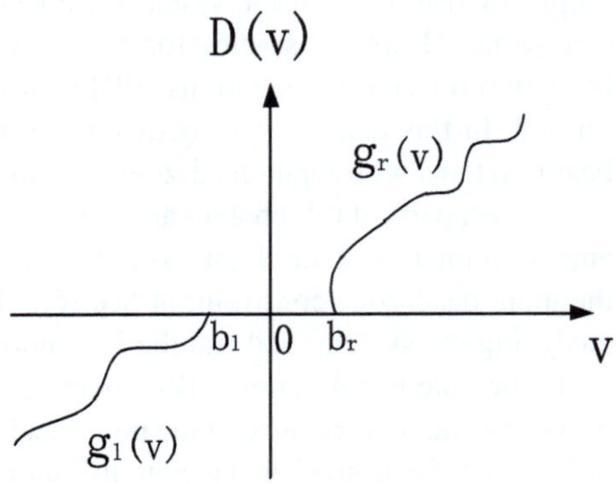

Fig. 6.2 The input dead-zone model

$$0 < k_{r0} < g'_r(v) < k_{r1}, v \in [b_r, +\infty) \tag{6.2}$$
$$0 < k_{l0} < g'_l(v) < k_{l1}, v \in (-\infty, b_l] \tag{6.3}$$

and $\beta_0 \leq \min\{k_{l0}, k_{r0}\}$ is a known positive constant, where $g'_r(v) = \frac{dg_r(z)}{dz}\big|_{z=v}$ and $g'_l(v) = \frac{dg_l(z)}{dz}\big|_{z=v}$.

For dealing with the nonlinearity issue of the dead-zone (6.1), we rewrote (6.1) as the expression below

$$D(v) = K^T(t)\Phi(t)v + d(v) \tag{6.4}$$

where the variables are defined as follows

$$\Phi(t) = [\varphi_r(t), \varphi_l(t)]^T \tag{6.5}$$

$$\varphi_r(t) = \begin{cases} 1, & v(t) > b_l \\ 0, & v(t) \leq b_l \end{cases} \tag{6.6}$$

$$\varphi_l(t) = \begin{cases} 1, & v(t) < b_r \\ 0, & v(t) \geq b_r \end{cases} \tag{6.7}$$

$$K(t) = [K_r(v(t)), K_l(v(t))]^T \tag{6.8}$$

$$K_r(v(t)) = \begin{cases} 0, & v(t) \leq b_l \\ g'_r(\xi_r(v(t))), & b_l < v(t) < +\infty \end{cases} \tag{6.9}$$

$$K_l(v(t)) = \begin{cases} g'_l(\xi_l(v(t))), & -\infty < v(t) < b_r \\ 0, & v(t) \geq b_r \end{cases} \tag{6.10}$$

$$d(v) = \begin{cases} -g'_r(\xi_r(v))b_r, & v \geq b_r \\ -[g'_r(\xi_r(v)) + g'_l(\xi_l(v))]v, & b_l < v < b_r \\ -g'_l(\xi_l(v))b_l, & v \leq b_l \end{cases} \tag{6.11}$$

It is remarked that $\xi_r(v) \in (b_r, v)$, if $b_r < v$; $\xi_r(v) \in (v, b_r)$, if $b_l < v \leq b_r$; $\xi_l(v) \in (v, b_l)$, if $v < b_l$; $\xi_l(v) \in (b_l, v)$, if $b_l \leq v < b_r$. From Assumption 6.3, we can also obtain that $K^T(t)\Phi(t) \geq \beta_0 > 0$.

Assumption 6.4 The energy of the unknown boundary external disturbance $\bar{d}(t)$ is bounded.

Remark 6.1 $|d(v)| \leq p^*$, where p^* is an unknown positive constant with $p^* = (k_{r1} + k_{l1})\max\{b_r, -b_l\}$.

Remark 6.2 The 'disturbance-like' term is defined as

$$d_1(t) = d(v) + \bar{d}(t) \tag{6.12}$$

which is also bounded within an unknown positive constant B, and B will be estimated in Sect. 6.3.

6.3 Control Design

In this section, we will discuss vibration control problem for flexible beam system. Our control objective is to make sure that, with input dead-zone nonlinearity $D(v)$ and boundary external disturbance $\bar{d}(t)$, the vibration of the flexible system can be suppressed and all the states of the closed-loop flexible mechanical systems are uniformly ultimately bounded.

We consider the following flexible Euler–Bernoulli beam system [5] in Fig. 6.1 as a more complex mechanical system with governing equation as

$$\rho\ddot{w}(x,t) + EIw''''(x,t) = Tw''(x,t) \tag{6.13}$$

$\forall(x,t) \in (0,L) \times [0,\infty)$, and boundary conditions as

$$w'(0,t) = 0 \tag{6.14}$$

$$w''(L,t) = 0 \tag{6.15}$$

$$w(0,t) = 0 \tag{6.16}$$

$$-EIw'''(L,t) + Tw'(L,t) = D(v) + \bar{d}(t) - M_s\ddot{w}(L,t) \tag{6.17}$$

$\forall t \in [0,\infty)$.

Based on the presentation of dead-zone (6.4), the boundary condition (6.15) can also be rewritten as

$$-EIw'''(L,t) + Tw'(L,t) = K^T(t)\Phi(t)v(t) + d_1(t) - M_s\ddot{w}(L,t) \tag{6.18}$$

where $d_1(t)$ bounded within an unknown positive constant B, and B will be estimated later. We design the desired controller $v(t)$ as

$$v(t) = \frac{1}{K^T(t)\Phi(t)}\Big[-EIw'''(L,t) + Tw'(L,t) - k_1M_s\dot{w}'(L,t) + k_2M_s\dot{w}'''(L,t) - ku_a(t) - \text{sgn}(u_a)\hat{B}(t)\Big] \tag{6.19}$$

where $u_a(t) = \dot{w}(L,t) + k_1w'(L,t) - k_2w'''(L,t)$ and $\hat{B}(t)$ is the observer of unknown constant B. We define the error of estimation as $\tilde{B}(t) = B - \hat{B}(t)$.

Therefore, $\dot{\tilde{B}}(t) = -\dot{\hat{B}}(t)$. The observer law is represented as

$$\dot{\hat{B}}(t) = |u_a| - \gamma \hat{B}(t) \tag{6.20}$$

We define the Lyapunov function candidate as

$$V_b(t) = V_{b1}(t) + V_{b2}(t) + \eta_b(t) \tag{6.21}$$

where

$$V_{b1}(t) = \frac{\beta k_2}{2}\rho \int_0^L [\dot{w}(x,t)]^2 dx + \frac{\beta k_2}{2} T \int_0^L [w'(x,t)]^2 dx + \frac{\beta k_2}{2} EI \int_0^L [w''(x,t)]^2 dx \tag{6.22}$$

$$V_{b2}(t) = \frac{1}{2} M_s u_a^2(t) + \frac{1}{2}\tilde{B}^2(t) \tag{6.23}$$

$$\eta_b(t) = \alpha\rho \int_0^L x\dot{w}(x,t)w'(x,t)dx \tag{6.24}$$

where α and β are two positive constants.

Lemma 6.1 *The Lyapunov function candidate given by* (6.21) *is upper and lower bounded as*

$$0 \leq \lambda_1(V_{b1}(t) + V_{b2}(t)) \leq V_b(t) \leq \lambda_2(V_{b1}(t) + V_{b2}(t)) \tag{6.25}$$

where λ_1 *and* λ_2 *are two positive constants.*

Proof Substituting (2.4) into (6.24) leads to

$$\begin{aligned} |\eta_b(t)| &\leq \alpha\rho L \int_0^L ([\dot{w}(x,t)]^2 + [w'(x,t)]^2)dx \\ &\leq \alpha_1 V_{b1}(t) \end{aligned} \tag{6.26}$$

where

$$\alpha_1 = \frac{2\alpha\rho L}{\min(k_2\beta\rho, k_2\beta T)} \tag{6.27}$$

Then, we have

$$-\alpha_1 V_{b1}(t) \leq \eta_b(t) \leq \alpha_1 V_{b1}(t) \tag{6.28}$$

Consider α satisfying $0 < \alpha < \frac{\min(k_2\beta\rho, k_2\beta T)}{2\rho L}$, we obtain

$$\alpha_2 = 1 - \alpha_1 = 1 - \frac{2\alpha\rho L}{\min(k_2\beta\rho, k_2\beta T)} > 0 \tag{6.29}$$

$$\alpha_3 = 1 + \alpha_1 = 1 + \frac{2\alpha\rho L}{\min(k_2\beta\rho, k_2\beta T)} > 1 \tag{6.30}$$

Then, we further have

$$0 \le \alpha_2 V_{b1}(t) \le V_{b1}(t) + \eta_b(t) \le \alpha_3 V_{b1}(t) \tag{6.31}$$

Given the Lyapunov functional candidate (6.21), we have

$$0 \le \lambda_1(V_{b1}(t) + V_{b2}(t)) \le V_b(t) \le \lambda_2(V_{b1}(t) + V_{b2}(t)) \tag{6.32}$$

where two positive constants $\lambda_1 = \min(\alpha_2, 1)$ and $\lambda_2 = \max(\alpha_3, 1)$ are given.

Lemma 6.2 *The time derivative of the Lyapunov function candidate* (6.21) *is upper bounded as*

$$\dot{V}_b(t) \le -\lambda V_b(t) + \varepsilon \tag{6.33}$$

where λ and ε are two positive constants.

Proof Differentiating (6.22) and substituting $u_a(t)$, we obtain

$$\begin{aligned}\dot{V}_{b1}(t) &= \beta T k_2 w'(L,t)\dot{w}(L,t) - \beta EI k_2 w'''(L,t)\dot{w}(L,t)\\ &= \frac{\beta EI}{2}u_a^2(t) - \frac{\beta EI}{2}\left\{[\dot{w}(L,t)]^2 + [k_1 w'(L,t)]^2 + [k_2 w'''(L,t)]^2\right\}\\ &\quad + \beta(Tk_2 - EIk_1)w'(L,t)\dot{w}(L,t) + \beta EI k_1 k_2 w'(L,t)w'''(L,t)\end{aligned} \tag{6.34}$$

Differentiating (6.23) and substituting the controller (6.19) and observer (6.20), we obtain

$$\begin{aligned}\dot{V}_{b2} &= M_s u_a \dot{u}_a + \tilde{B}(t)\dot{\tilde{B}}(t)\\ &= u_a\Big[EIw'''(L,t) - Tw'(L,t) + k_1 M_s \dot{w}'(L,t) - k_2 M_s \dot{w}'''(L,t)\\ &\quad + K^T(t)\Phi(t)v(t) + d_1(t)\Big] - \tilde{B}(t)\dot{\hat{B}}(t)\\ &= -ku_a^2 - |u_a|\hat{B}(t) + d_1(t)u_a - |u_a|\tilde{B}(t) + \gamma\tilde{B}(t)\hat{B}(t)\\ &\le -ku_a^2 - |u_a|(B - |d_1(t)|) - \frac{\gamma}{2}\tilde{B}^2(t) + \frac{\gamma}{2}B^2\\ &\le -ku_a^2 - \frac{\gamma}{2}\tilde{B}^2(t) + \frac{\gamma}{2}B^2\end{aligned} \tag{6.35}$$

Differentiating (6.24), we obtain

$$\begin{aligned}\dot{\eta}_b(t) &= -\alpha EILw'(L,t)w'''(L,t) - \frac{3\alpha EI}{2}\int_0^L [w''(x,t)]^2 dx + \frac{\alpha TL}{2}[w'(L,t)]^2\\ &\quad - \frac{\alpha T}{2}\int_0^L [w'(x,t)]^2 dx + \frac{\alpha\rho L}{2}[\dot{w}(L,t)]^2 - \frac{\alpha\rho}{2}\int_0^L [\dot{w}(x,t)]^2 dx\end{aligned} \tag{6.36}$$

Combining (6.34)–(6.36), we have

$$
\begin{aligned}
\dot{V}_b(t) \leq & -\frac{\alpha\rho}{2}\int_0^L [\dot{w}(x,t)]^2 dx - \frac{\alpha T}{2}\int_0^L [w'(x,t)]^2 dx - \left(k - \frac{\beta EI}{2}\right)u_a^2 - \frac{\gamma}{2}\tilde{B}^2(t) \\
& + \beta(Tk_2 - EIk_1)w'(L,t)\dot{w}(L,t) + (\beta EIk_1k_2 - \alpha EIL)w'(L,t)w'''(L,t) \\
& - \left(\frac{\beta EI}{2} - \frac{\alpha\rho L}{2}\right)[\dot{w}(L,t)]^2 - \left(\frac{\beta EIk_1^2}{2} - \frac{\alpha TL}{2}\right)[w'(L,t)]^2 \\
& - \frac{\beta EIk_2^2}{2}[w'''(L,t)]^2 - \frac{3\alpha EI}{2}\int_0^L [w''(x,t)]^2 dx + \frac{\gamma}{2}B^2
\end{aligned} \tag{6.37}
$$

Applying inequalities in Lemma 2.1, we have

$$
\begin{aligned}
\dot{V}_b(t) \leq & -\frac{\alpha\rho}{2}\int_0^L [\dot{w}(x,t)]^2 dx - \frac{\alpha T}{2}\int_0^L [w'(x,t)]^2 dx - \left(k - \frac{\beta EI}{2}\right)u_a^2 - \frac{\gamma}{2}\tilde{B}^2(t) \\
& - \left(\frac{\beta EI}{2} - \frac{\alpha\rho L}{2} - \beta|Tk_2 - EIk_1|\delta_1\right)[\dot{w}(L,t)]^2 \\
& - \left(\frac{\beta EIk_2^2}{2} - |\beta EIk_1k_2 - \alpha EIL|\delta_2\right)[w'''(L,t)]^2 \\
& - \left(\frac{\beta EIk_1^2}{2} - \frac{\alpha TL}{2} - \frac{\beta|Tk_2 - EIk_1|}{\delta_1} - \frac{|\beta EIk_1k_2 - \alpha EIL|}{\delta_2}\right)[w'(L,t)]^2 \\
& - \frac{3\alpha EI}{2}\int_0^L [w''(x,t)]^2 dx + \frac{\gamma}{2}B^2 \\
\leq & -\lambda_3(V_{b1}(t) + V_{b2}(t)) + \varepsilon
\end{aligned} \tag{6.38}
$$

where δ_1 and δ_2 are two positive constants, $\lambda_3 = \min\left\{\frac{\alpha}{k_2\beta}, \frac{2k-\beta EI}{M_s}, \gamma\right\} > 0$ and $\varepsilon = \frac{\gamma}{2}B^2 > 0$. Choosing $k > \frac{\beta EI}{2}$, and the following conditions should be satisfied

$$\frac{\beta EI}{2} - \frac{\alpha\rho L}{2} - \beta|Tk_2 - EIk_1|\delta_1 \geq 0 \tag{6.39}$$

$$\frac{\beta EIk_2^2}{2} - |\beta EIk_1k_2 - \alpha EIL|\delta_2 \geq 0 \tag{6.40}$$

$$\frac{\beta EIk_1^2}{2} - \frac{\alpha TL}{2} - \frac{\beta|Tk_2 - EIk_1|}{\delta_1} - \frac{|\beta EIk_1k_2 - \alpha EIL|}{\delta_2} \geq 0 \tag{6.41}$$

Combining Inqes. (6.32) and (6.38), we obtain

$$\dot{V}_b(t) \leq -\lambda V_b(t) + \varepsilon \tag{6.42}$$

where $\lambda = \lambda_3/\lambda_2$.

Theorem 6.1 *For the system dynamics of beam described by* (6.13) *and boundary conditions* (6.14)–(6.17)*, with the proposed control* (6.19)*, dead-zone nonlinearity* (6.13) *and observer* (6.20)*, for bounded initial conditions, if the control parameters fulfill previously determined conditions, then the closed-loop system signal* $w(x,t)$ *is stable.*

6.4 Numerical Simulations

In order to verify the effectiveness of the proposed control method, the finite difference method is applied in the simulation. The initial conditions of the system are given as: $w(x,0) = \frac{x}{L}$, $\dot{w}(x,0) = 0$ and the corresponding boundary disturbance on the tip payload is defined as: $\bar{d}(t) = 1 + \sin(2\pi t) + \cos(3\pi t)$. And the parameters of the systems are given in Table 6.1.

Table 6.1 Parameters of the flexible beam system

Parameters	Description	Value
L	Length of the flexible beam	1 m
T	Tension of the flexible beam	10 N
ρ	Uniform mass per unit of the flexible beam	0.5 kg/m
M_s	Mass of the tip payload	1 kg
EI	Bending stiffness of the flexible beam	5 Nm^2

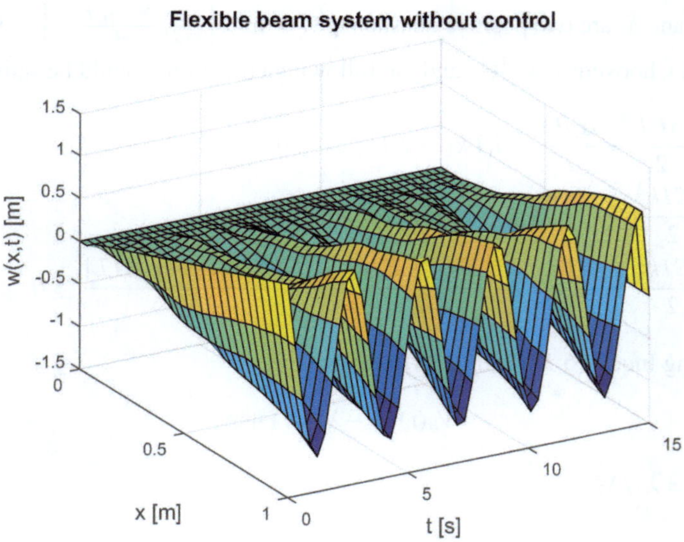

Fig. 6.3 Displacement of the flexible beam system without control

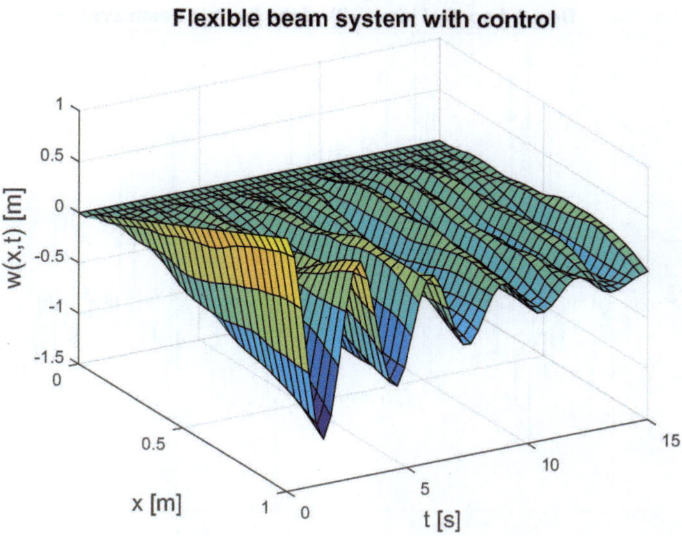

Fig. 6.4 Displacement of the flexible beam system with control

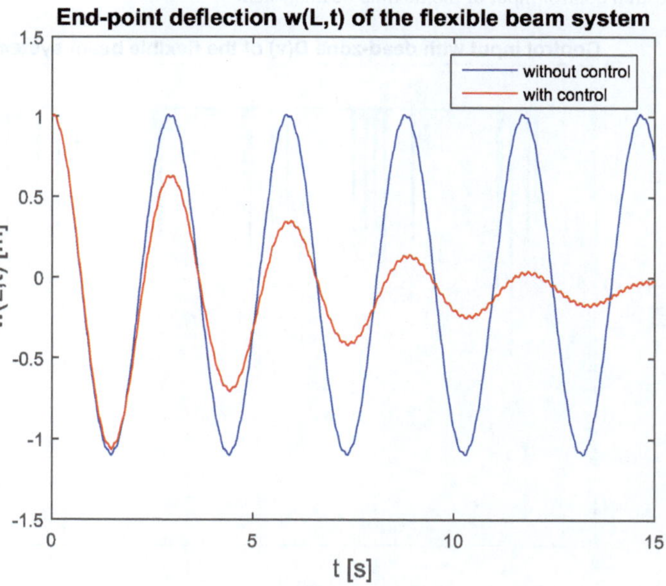

Fig. 6.5 End-point deflection of the flexible beam system

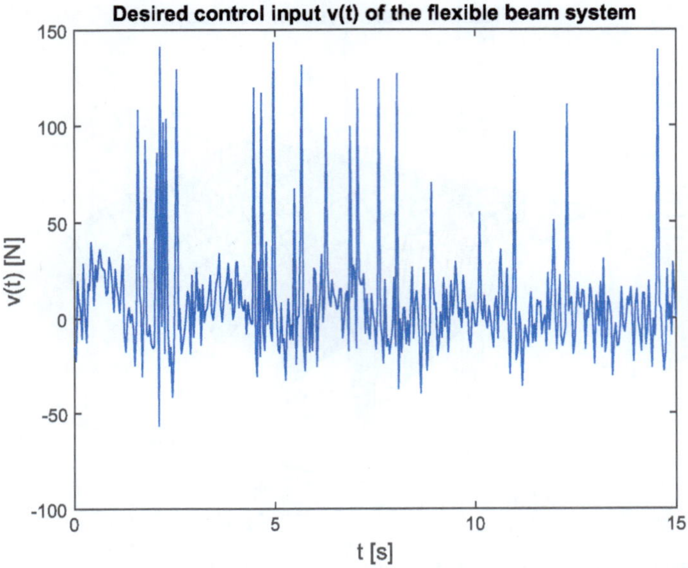

Fig. 6.6 Desired control input of the flexible beam system

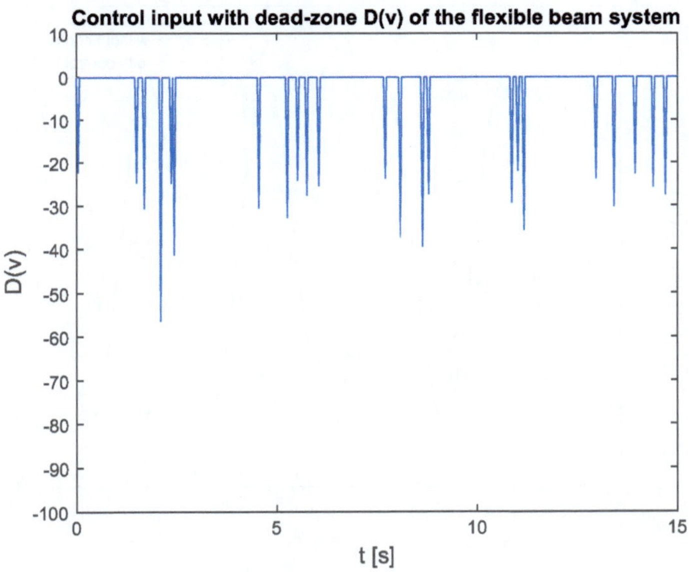

Fig. 6.7 Control input with dead-zone of the flexible beam system

In this section, the simulations of the flexible beam system are discussed in the following cases (Figs. 6.3, 6.4, 6.5, 6.6 and 6.7):

(i) the systems are freely vibrated without control;
(ii) the systems are performed with the proposed control method.

6.5 Conclusion

In this chapter, boundary control architecture has been proposed for continuous-time PDE systems preceded by input dead-zone nonlinearity. By using a new description of the dead-zone, boundary control schemes have been developed to regulate the deformation of Euler–Bernoulli beam system even in the presence of the external disturbance. It has been proved that the proposed control law can ensure uniformly ultimately boundness of the entire system and achieve stabilization. Simulations have been performed on beam systems to illustrate and clarify the effectiveness of the proposed control approach.

References

1. X.-S. Wang, C.-Y. Su, H. Hong, Robust adaptive control of a class of nonlinear systems with unknown dead-zone. Automatica **40**(3), 407–413 (2004)
2. X.-S. Wang, H. Hong, C.-Y. Su, Model reference adaptive control of continuous-time systems with an unknown input dead-zone. IEE Proc. Control Theory Appl. **150**(3), 261–266 (2003)
3. M. Chen, S.S. Ge, B. How, Robust adaptive neural network control for a class of uncertain MIMO nonlinear systems with input nonlinearities. IEEE Trans. Neural Netw. **21**(5), 796–812 (2010)
4. T.-P. Zhang, S.S. Ge, Adaptive dynamic surface control of nonlinear systems with unknown dead zone in pure feedback form. Automatica **44**(7), 1895–1903 (2008)
5. S.S. Ge, S. Zhang, W. He, Vibration control of an Euler-Bernoulli beam under unknown spatiotemporally varying disturbance. Int. J. Control **84**(5), 947–960 (2011)

Chapter 7
Vibration Control of a Flexible Beam with Input Backlash

7.1 Introduction

In this chapter, the nonsmooth input backlash is taken into consideration. The input backlash will be caused by the gears which can control the motor to realize the proposed actual input. The space between gears results in a backlash function which relates to the input and output angles. If backlash effect in gears is not reduced, it will degrade the system's performance, reduce control accuracy and even make the system unstable. Some solutions have been developed to reduce the effect of backlash in the aspects of mechanical. However, they will decrease the accuracy of the system. In addition, they are expensive, energy consuming usually and would increase the weight of the system. Therefore, some works research the theoretic method to overcome these drawbacks. In [1], a robust adaptive control is designed for nonlinear dynamical systems with a backlash-like hysteresis. In [2], the authors present sliding mode controllers to analyze the stability of a nonlinear uncertain plant with dead-zone or backlash in the actuator devices.

In this chapter, we focus on the vibration control problem for a flexible beam with input backlash. The dynamics of the beam are represented by a PDE and four ODEs involving functions of both space and time. Many effective methods are developed for nonlinear uncertain systems with dead-zone or backlash, such as [1–6]. However, most of the previous works on dealing with the input nonlinearity are investigated for the systems described by ODEs. There is little information about how to handle the backlash nonlinearity for PDEs and there is a need to explore an effective method for the control of flexible structures with input backlash. To the best of our knowledge, the result is the first complete solution of boundary control to a flexible beam system with input backlash.

In this chapter, an active boundary control is designed to reduce the vibration of the flexible beam under the time-varying external disturbance based on a linear input which is transformed from the input backlash by choosing an optimal method.

The rest content of this chapter are organized as follows. The preliminaries of the chapter is presented for the subsequent development and the nonsmooth nonlinear

W. He and J. Liu, *Active Vibration Control and Stability Analysis of Flexible Beam Systems*, https://doi.org/10.1007/978-981-10-7539-1_7

input backlash is transformed to linear input in Sect. 7.2. In Sect. 7.3, the boundary control law and disturbance observer are designed via Lyapunov's direct method to analyze the stability of the flexible beam system. The effectiveness of the proposed control method is proved by the performance of the numerical simulation with finite difference method in Sect. 7.4. The conclusion of the chapter is given in Sect. 7.5.

7.2 Problem Formulation

For the convenience of stability analysis, we present the following lemmas for the subsequent development.

In [7], the authors proposed the following backlash nonlinearity shown in Fig. 7.1 as

$$u(t) = B(v) = \begin{cases} m(v(t) - B_r), & \text{if} \quad \dot{v} > 0 \quad \text{and} \quad u(t) = m(v(t) - B_r) \\ m(v(t) - B_l), & \text{if} \quad \dot{v} < 0 \quad \text{and} \quad u(t) = m(v(t) - B_l) \\ u(t\_), & \text{otherwise} \end{cases} \tag{7.1}$$

where m is the constant slope of the lines, B_r and B_l are constant parameters. $u(t\_)$ means no change occurs in $u(t)$ which is the boundary control input working on the flexible beam system.

Backlash is widely presented using two parallel slanted lines which are connected with some parallel horizontal line segments. The parallel slanted lines are active when the input signal is changing in the same direction. The inner parallel horizontal line segments indicate the gap between the parallel slanted lines, therefore, the output will change from a parallel slanted line to another slanted line when $\dot{v}(t)$ changes.

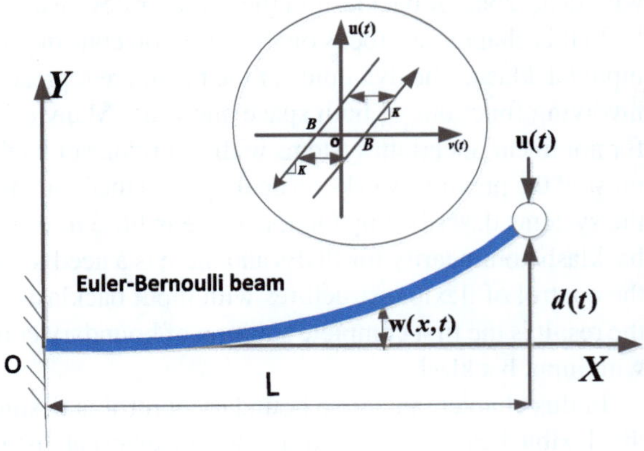

Fig. 7.1 Diagram of backlash

Several assumptions are also given as follows

Assumption 7.1 The backlash output $B(v)$ is not available for measurement. The backlash parameters, m, B_r and B_l, are unknown bounded constants, but their signs are known, i.e., $m > 0$, $B_r > 0$ and $B_l < 0$.

Assumption 7.2 The backlash parameters are within known bounded

$$0 < m_{\min} \leq m \leq m_{\max} \tag{7.2}$$
$$0 < (mB_r)_{\min} \leq mB_r \leq (mB_r)_{\max} \tag{7.3}$$
$$(mB_l)_{\min} \leq mB_l \leq (mB_l)_{\max} < 0 \tag{7.4}$$

For dealing with the nonlinearity issue of the backlash [7], we rewrite (7.1) as follows expression

$$u(t) = B(v) = mv(t) + d_0(v) \tag{7.5}$$

where $d_0(v)$ is the nonlinearity error for input backlash and can be written as

$$d_0(v) = \begin{cases} -mB_r, & \text{if} \quad \dot{v} > 0 \quad \text{and} \quad u(t) = m(v(t) - B_r) \\ -mB_l, & \text{if} \quad \dot{v} < 0 \quad \text{and} \quad u(t) = m(v(t) - B_l) \\ u(t\_) - mv(t), & \text{otherwise} \end{cases} \tag{7.6}$$

Assumption 7.3 We assume that the unknown boundary external disturbance $d(t)$ is bounded. This is a reasonable assumption as the time-varying disturbance $d(t)$ has finite energy and hence is bounded.

Remark 7.1 $|d_0(v)| \leq p^*$, where p^* is an unknown positive constant with $p^* = \max\{(mB_r)_{\max}, -(mB_l)_{\min}\}$.

Remark 7.2 The 'disturbance-like' term is defined as

$$d_1(t) = d_0(v) + d(t) \tag{7.7}$$

which is also bounded within an unknown positive constant D, and D will be estimated in Sect. 7.3.

7.3 Control Design

We consider the following flexible Euler–Bernoulli beam system for the small diameter-to-length rate [8] as a more complicated model with governing equation as

$$\rho\ddot{w}(x,t) + EIw''''(x,t) = Tw''(x,t) \tag{7.8}$$

$\forall(x,t) \in (0,L) \times [0,\infty)$, and boundary conditions as

$$w'(0,t) = 0 \tag{7.9}$$

$$w''(L,t) = 0 \tag{7.10}$$

$$w(0,t) = 0 \tag{7.11}$$

$$-EIw'''(L,t) + Tw'(L,t) = u(t) + d(t) - M_s\ddot{w}(L,t) \tag{7.12}$$

$\forall t \in [0,\infty)$. Based on the presentation of backlash (7.5), the boundary condition (7.12) can also be rewritten as

$$-EIw'''(L,t) + Tw'(L,t) = mv(t) + d_1(t) - M_s\ddot{w}(L,t) \tag{7.13}$$

where $d_1(t)$ is bounded within an unknown positive constant D, and D will be estimated later.

In this section, we will design the desired controller $v(t)$, and the boundary control is designed to make sure that, with input backlash nonlinearity $B(v)$ and boundary external disturbance $d(t)$, all the states of the closed-loop flexible structures are uniformly ultimately bounded. The desired controller is designed as

$$v(t) = \frac{1}{m}\Big[-EIw'''(L,t) + Tw'(L,t) - k_1M_s\dot{w}'(L,t) + k_2M_s\dot{w}'''(L,t) - ku_a(t) - \text{sgn}(u_a(t))\hat{D}(t)\Big] \tag{7.14}$$

where $u_a(t) = \dot{w}(L,t) + k_1w'(L,t) - k_2w'''(L,t)$ and $\hat{D}(t)$ is the observer of unknown constant D, k_1 and k_2 are two positive control gains. We define the error of estimation as $\tilde{D}(t) = D - \hat{D}(t)$. Therefore, $\dot{\tilde{D}}(t) = -\dot{\hat{D}}(t)$. The observer law is represented as

$$\dot{\hat{D}}(t) = |u_a(t)| - \gamma\hat{D}(t) \tag{7.15}$$

Assumption 7.4 For the system dynamics described by (7.8)–(7.12), with the boundary control (7.14), we assume that the closed-loop system is well posed.

We define the Lyapunov candidate function as

$$V_b(t) = V_{b1}(t) + V_{b2}(t) + \eta_b(t) \tag{7.16}$$

where

$$V_{b1}(t) = \frac{\beta k_2}{2}\rho\int_0^L[\dot{w}(x,t)]^2dx + \frac{\beta k_2}{2}T\int_0^L[w'(x,t)]^2dx + \frac{\beta k_2}{2}EI\int_0^L[w''(x,t)]^2dx \tag{7.17}$$

$$V_{b2}(t) = \frac{1}{2}M_su_a^2(t) + \frac{1}{2}\tilde{D}^2(t) \tag{7.18}$$

$$\eta_b(t) = \alpha\rho \int_0^L x\dot{w}(x,t)w'(x,t)dx \tag{7.19}$$

where α and β are two positive constants which will be chosen later.

Lemma 7.1 *The Lyapunov candidate function* (7.16) *is bounded as*

$$0 \le \lambda_1(V_{b1}(t) + V_{b2}(t)) \le V_b(t) \le \lambda_2(V_{b1}(t) + V_{b2}(t)) \tag{7.20}$$

where λ_1 and λ_2 are two positive constants.

Proof From definition of $V_{b1}(t)$, we know that $V_{b1}(t)$ is bounded as

$$\begin{aligned} V_{b1}(t) &= \frac{\beta k_2}{2}\rho \int_0^L [\dot{w}(x,t)]^2 dx + \frac{\beta k_2}{2}T \int_0^L [w'(x,t)]^2 dx + \frac{\beta k_2}{2}EI \int_0^L [w''(x,t)]^2 dx \\ &\ge \frac{\beta k_2}{2}\rho \int_0^L [\dot{w}(x,t)]^2 dx + \frac{\beta k_2}{2}T \int_0^L [w'(x,t)]^2 dx \\ &\ge \frac{\min(\beta k_2\rho, \beta k_2 T)}{2}\left(\int_0^L [\dot{w}(x,t)]^2 dx + \int_0^L [w'(x,t)]^2 dx\right) \end{aligned} \tag{7.21}$$

Since $x \in [0, L]$, and substituting Lemma 2.1 into (7.19) leads to

$$\begin{aligned} |\eta_b(t)| &\le \alpha\rho \int_0^L x([\dot{w}(x,t)]^2 + [w'(x,t)]^2)dx \\ &\le \alpha\rho L \int_0^L ([\dot{w}(x,t)]^2 + [w'(x,t)]^2)dx \\ &\le \frac{2\alpha\rho L}{\min(\beta k_2\rho, \beta k_2 T)} V_{b1}(t) \\ &= \alpha_1 V_{b1}(t) \end{aligned} \tag{7.22}$$

where

$$\alpha_1 = \frac{2\alpha\rho L}{\min(\beta k_2\rho, \beta k_2 T)} \tag{7.23}$$

Then, we have

$$-\alpha_1 V_{b1}(t) \le \eta_b(t) \le \alpha_1 V_{b1}(t) \tag{7.24}$$

Consider α is a small positive weighting constant satisfying $0 < \alpha < \frac{\min(\beta k_2\rho, \beta k_2 T)}{2\rho L}$, we obtain

$$\alpha_2 = 1 - \alpha_1 = 1 - \frac{2\alpha\rho L}{\min(\beta k_2\rho, \beta k_2 T)} > 0 \tag{7.25}$$

$$\alpha_3 = 1 + \alpha_1 = 1 + \frac{2\alpha\rho L}{\min(\beta k_2\rho, \beta k_2 T)} > 1 \tag{7.26}$$

Then, we further have

$$0 \leq \alpha_2 V_{b1}(t) \leq V_{b1}(t) + \eta_b(t) \leq \alpha_3 V_{b1}(t) \tag{7.27}$$

Given the Lyapunov candidate function (7.16), we have

$$0 \leq \lambda_1(V_{b1}(t) + V_{b2}(t)) \leq V_b(t) \leq \lambda_2(V_{b1}(t) + V_{b2}(t)) \tag{7.28}$$

where two positive constants $\lambda_1 = \min(\alpha_2, 1) = \alpha_2$ and $\lambda_2 = \max(\alpha_3, 1) = \alpha_3$ are given.

Lemma 7.2 *The time derivative of the Lyapunov candidate function* (7.16) *is bounded as*

$$\dot{V}_b(t) \leq -\lambda V_b(t) + \varepsilon \tag{7.29}$$

where λ *and* ε *are two positive constants.*

Proof Differentiating (7.17), we have

$$\begin{aligned}\dot{V}_{b1}(t) &= \beta k_2 \rho \int_0^L \dot{w}(x,t)\ddot{w}(x,t)dx + \beta k_2 T \int_0^L w'(x,t)\dot{w}'(x,t)dx \\ &+ \beta k_2 EI \int_0^L w''(x,t)\dot{w}''(x,t)dx\end{aligned} \tag{7.30}$$

Substituting the governing equation (7.8), we have

$$\begin{aligned}\dot{V}_{b1}(t) &= \beta k_2 \int_0^L \dot{w}(x,t)\left[Tw''(x,t) - EIw''''(x,t)\right]dx + \beta k_2 T \int_0^L w'(x,t)\dot{w}'(x,t)dx \\ &+ \beta k_2 EI \int_0^L w''(x,t)\dot{w}''(x,t)dx \\ &= A_1(t) + A_2(t)\end{aligned} \tag{7.31}$$

where

$$A_1(t) = \beta k_2 T \int_0^L \dot{w}(x,t)w''(x,t)dx + \beta k_2 T \int_0^L w'(x,t)\dot{w}'(x,t)dx \tag{7.32}$$

$$A_2(t) = -\beta k_2 EI \int_0^L \dot{w}(x,t)w''''(x,t)dx + \beta k_2 EI \int_0^L w''(x,t)\dot{w}''(x,t)dx \tag{7.33}$$

Using integration by parts and boundary condition (7.9), we obtain $A_1(t)$ as

$$
\begin{aligned}
A_1(t) &= \beta k_2 T \int_0^L \dot{w}(x,t) w''(x,t) dx + \beta k_2 T \int_0^L w'(x,t) \dot{w}'(x,t) dx \\
&= \beta k_2 T \dot{w}(x,t) w'(x,t) \Big|_0^L - \beta k_2 T \int_0^L w'(x,t) \dot{w}'(x,t) dx + \beta k_2 T \int_0^L w'(x,t) \dot{w}'(x,t) dx \\
&= \beta k_2 T \dot{w}(L,t) w'(L,t)
\end{aligned} \tag{7.34}
$$

In a similar manner, we obtain $A_2(t)$ as

$$
\begin{aligned}
A_2(t) =& -\beta k_2 EI \int_0^L \dot{w}(x,t) w''''(x,t) dx + \beta k_2 EI \int_0^L w''(x,t) \dot{w}''(x,t) dx \\
=& -\beta k_2 EI \dot{w}(x,t) w'''(x,t) \Big|_0^L + \beta k_2 EI \int_0^L \dot{w}'(x,t) w'''(x,t) dx \\
&+ \beta k_2 EI \int_0^L w''(x,t) \dot{w}''(x,t) dx \\
=& -\beta k_2 EI \dot{w}(x,t) w'''(x,t) \Big|_0^L + \beta k_2 EI \dot{w}'(x,t) w''(x,t) \Big|_0^L
\end{aligned} \tag{7.35}
$$

From boundary conditions (7.9)–(7.11), we have

$$
A_2(t) = -\beta k_2 EI \dot{w}(L,t) w'''(L,t) \tag{7.36}
$$

Combining $A_1(t)$ and $A_2(t)$, we have $\dot{V}_{b1}(t)$ as

$$
\dot{V}_{b1}(t) = \beta k_2 T \dot{w}(L,t) w'(L,t) - \beta k_2 EI \dot{w}(L,t) w'''(L,t) \tag{7.37}
$$

Recall that $u_a(t) = \dot{w}(L,t) + k_1 w'(L,t) - k_2 w'''(L,t)$, then the last term in (7.37) is written as

$$
\begin{aligned}
-\beta k_2 EI \dot{w}(L,t) w'''(L,t) =& \frac{\beta EI}{2} u_a^2(t) - \frac{\beta EI}{2} \left\{ [\dot{w}(L,t)]^2 + [k_1 w'(L,t)]^2 + [k_2 w'''(L,t)]^2 \right\} \\
& - \beta EI k_1 w'(L,t) \dot{w}(L,t) + \beta EI k_1 k_2 w'(L,t) w'''(L,t)
\end{aligned} \tag{7.38}
$$

Then, we obtain

$$
\begin{aligned}
\dot{V}_{b1}(t) =& \frac{\beta EI}{2} u_a^2(t) - \frac{\beta EI}{2} \left\{ [\dot{w}(L,t)]^2 + [k_1 w'(L,t)]^2 + [k_2 w'''(L,t)]^2 \right\} \\
& + \beta (T k_2 - EI k_1) w'(L,t) \dot{w}(L,t) + \beta EI k_1 k_2 w'(L,t) w'''(L,t)
\end{aligned} \tag{7.39}
$$

Differentiating (7.18) and substituting boundary condition (7.12), then, using the controller (7.14) and observer (7.15), we obtain

$$
\begin{aligned}
\dot{V}_{b2} =& M_s u_a(t) \dot{u}_a(t) + \tilde{D}(t) \dot{\tilde{D}}(t) \\
\leq& -k u_a^2(t) - \frac{\gamma}{2} \tilde{D}^2(t) + \frac{\gamma}{2} D^2
\end{aligned} \tag{7.40}
$$

Differentiating (7.19) and substituting the governing equation (7.8) and boundary conditions (7.9)–(7.11), we obtain

$$\begin{aligned}\dot{\eta}_b(t) =& -\alpha EILw'(L,t)w'''(L,t) - \frac{3\alpha EI}{2}\int_0^L [w''(x,t)]^2 dx + \frac{\alpha TL}{2}[w'(L,t)]^2 \\ &- \frac{\alpha T}{2}\int_0^L [w'(x,t)]^2 dx + \frac{\alpha\rho L}{2}[\dot{w}(L,t)]^2 - \frac{\alpha\rho}{2}\int_0^L [\dot{w}(x,t)]^2 dx\end{aligned} \tag{7.41}$$

Combining (7.38)–(7.40), and applying inequalities in Lemma 2.2, we have

$$\begin{aligned}\dot{V}_b(t) \leq& -\frac{\alpha\rho}{2}\int_0^L [\dot{w}(x,t)]^2 dx - \frac{\alpha T}{2}\int_0^L [w'(x,t)]^2 dx - \left(k - \frac{\beta EI}{2}\right)u_a^2 - \frac{\gamma}{2}\tilde{D}^2(t) \\ &- \left(\frac{\beta EI}{2} - \frac{\alpha\rho L}{2} - \beta|Tk_2 - EIk_1|\delta_1\right)[\dot{w}(L,t)]^2 - \frac{3\alpha EI}{2}\int_0^L [w''(x,t)]^2 dx \\ &- \left(\frac{\beta EIk_2^2}{2} - |\beta EIk_1k_2 - \alpha EIL|\delta_2\right)[w'''(L,t)]^2 + \frac{\gamma}{2}D^2 \\ &- \left(\frac{\beta EIk_1^2}{2} - \frac{\alpha TL}{2} - \frac{\beta|Tk_2 - EIk_1|}{\delta_1} - \frac{|\beta EIk_1k_2 - \alpha EIL|}{\delta_2}\right)[w'(L,t)]^2 \\ \leq& -\lambda_3(V_{b1}(t) + V_{b2}(t)) + \varepsilon\end{aligned} \tag{7.42}$$

where δ_1 and δ_2 are two positive constants, and $\varepsilon = \frac{\gamma}{2}D^2 > 0$. Choosing $k - \frac{\beta EI}{2} > 0$, then $k_1, k_2, \alpha, \beta, \delta_1$ and δ_2 are designed to satisfy the following conditions

$$\frac{\beta EI}{2} - \frac{\alpha\rho L}{2} - \beta|Tk_2 - EIk_1|\delta_1 \geq 0 \tag{7.43}$$

$$\frac{\beta EIk_2^2}{2} - |\beta EIk_1k_2 - \alpha EIL|\delta_2 \geq 0 \tag{7.44}$$

$$\frac{\beta EIk_1^2}{2} - \frac{\alpha TL}{2} - \frac{\beta|Tk_2 - EIk_1|}{\delta_1} - \frac{|\beta EIk_1k_2 - \alpha EIL|}{\delta_2} \geq 0 \tag{7.45}$$

$$\lambda_3 = \min\left\{\frac{\alpha}{\beta k_2}, \frac{2k - \beta EI}{M_s}, \frac{3\alpha}{\beta k_2}, \gamma\right\} > 0 \tag{7.46}$$

Combining (7.28) and (7.41), we obtain

$$\dot{V}_b(t) \leq -\lambda V_b(t) + \varepsilon \tag{7.47}$$

where $\lambda = \lambda_3/\lambda_2$.

The theorem is obtained to reveal the stability of the closed-loop system with input backlash nonlinearity $B(v)$ and boundary external disturbance $\bar{d}(t)$ using the above lemmas.

Theorem 7.1 *For the system dynamics of beam described by* (7.8) *and boundary conditions* (7.9)–(7.12), *with the proposed control* (7.14), *backlash nonlinearity* (7.1) *and observer* (7.15), *for bounded initial conditions, if the control parameters fulfill* *previously determined conditions, then the closed-loop system signal $w(x,t)$ is stable.*

Proof Integrating (7.47), we obtain

$$V_b(t) \leq V_b(0)e^{-\lambda t} + \frac{\varepsilon}{\lambda} \in \mathcal{L}_\infty \tag{7.48}$$

which implies $V_b(t)$ is bounded. Utilizing Lemma 2.4, we have

$$\frac{\beta k_2}{2L} T w^2(x,t) \leq \frac{\beta k_2}{2} T \int_0^L [w'(x,t)]^2 dx \leq V_{b1}(t) \leq \frac{1}{\lambda_1} V_b(t) \in \mathcal{L}_\infty \tag{7.49}$$

Appropriately rearranging the terms of the above inequality, we obtain $w(x,t)$ is stable as follows

$$|w(x,t)| \leq \sqrt{\frac{2L}{\beta k_2 T \lambda_1}\left(V_b(0)e^{-\lambda t} + \frac{\varepsilon}{\lambda}\right)}, \forall (x,t) \in [0,L] \times [0,\infty) \tag{7.50}$$

Furthermore, when t tends to infinity, we have

$$\lim_{t\to\infty} |w(x,t)| \leq \sqrt{\frac{2L\varepsilon}{\beta k_2 T \lambda_1 \lambda}}, \quad \forall x \in [0,L] \tag{7.51}$$

From (7.51), it can be seen that the displacement $w(x,t)$ converges to a small positive constant at the rate of convergence λ as $t \to \infty$.

Remark 7.3 Please note that, from (7.51), it is clear that the steady system state $w(x,t)$ can be made small provided that the design control parameters are appropriately selected. It can be seen that the increase in the control parameters k and γ or decrease in the control parameter k_2 will result in a larger λ_3. Then, the value of λ will increase and produce a better vibration suppression performance. However, increasing k will bring a high gain control scheme. Therefore, in practical applications, the design parameters should be adjusted carefully for achieving suitable transient performance and control action.

Remark 7.4 The signals which make up the proposed boundary control law can be measured directly by sensors or be obtained using a backward difference algorithm. In this control law, the displacement $w(L,t)$ can be measured by a laser displacement sensor at the top boundary of the beam, $w'(L,t)$ can be obtained by an inclinometer and $w'''(L,t)$ can be measured with a shear force sensor. In addition, $\dot{w}(L,t)$, $\dot{w}'(L,t)$ and $\dot{w}'''(L,t)$ can be obtained with a backward difference algorithm based on the measured values. It needs to be noted that differentiating twice and thrice the position $w(L,t)$ with respect to time to obtain $\ddot{w}(L,t)$ and $\dddot{w}(L,t)$ respectively, are undesirable in practice since the existing noise will be amplified.

Table 7.1 Parameters of the flexible beam

Parameter	Description	Value
L	Length of the beam	1 m
M_s	Mass of the tip payload	6 kg
T	Tension of the beam	2 N
EI	Bending stiffness of the beam	8 Nm^2
ρ	Mass per unit length of the beam	2 kg/m

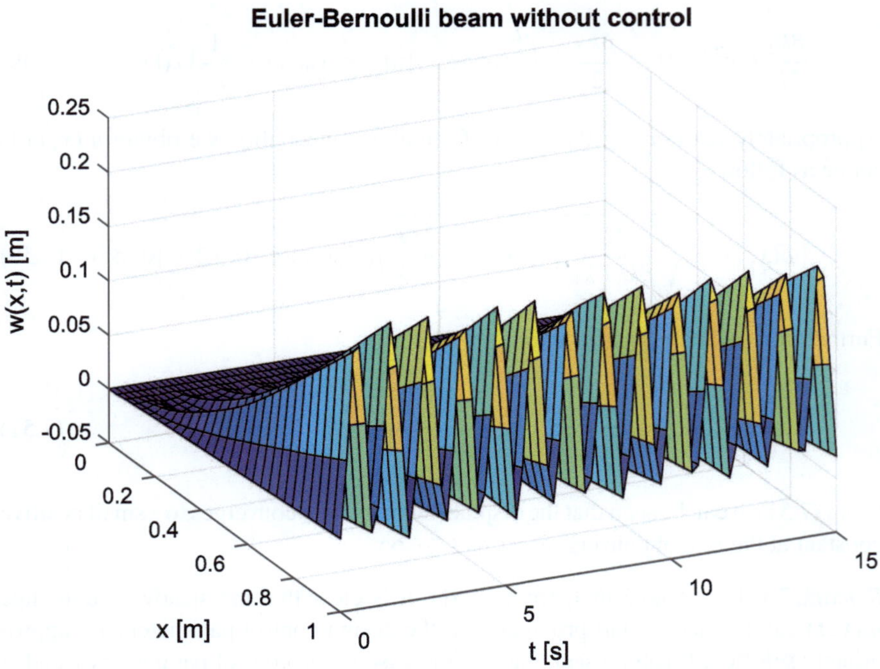

Fig. 7.2 Displacement of the flexible beam system without control

7.4 Numerical Simulations

In order to verity the effectiveness of the proposed boundary control law for the flexible beam system with input backlash characteristic, we make the numerical simulation and get satisfying performance. For the flexible beam system, the parameters of the system are given in the Table 7.1.

The flexible beam system is subject to the external vary-time disturbances, and in the previous research there are some models to approximate the real disturbances environment to analyze the effects of the disturbances acting on this system. In this section, the environment disturbance is expressed as $\bar{d}(t) = 1 + \sin(2\pi t) + \cos(3\pi t)$ to simulate the real environmental disturbance [8].

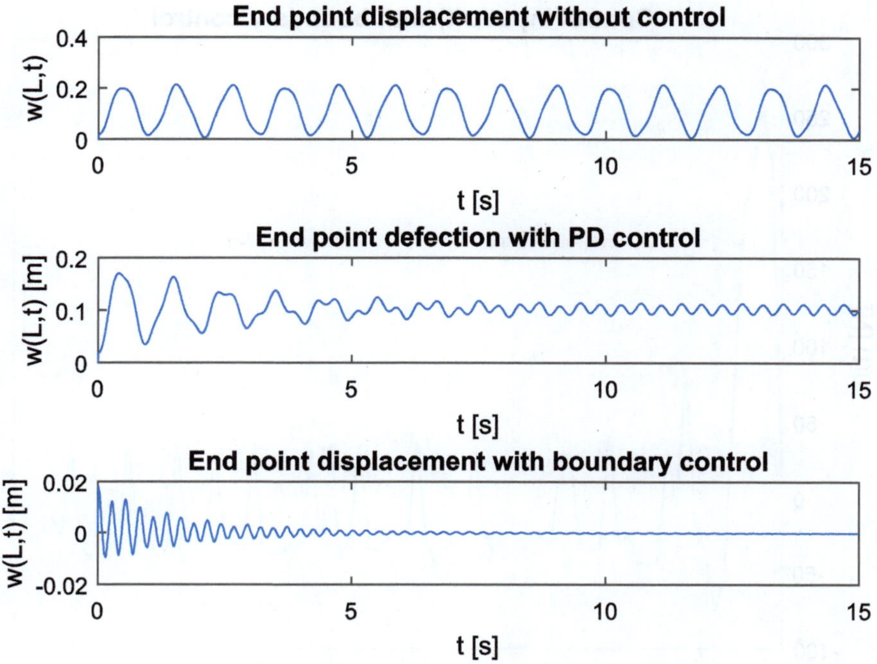

Fig. 7.5 End-point deflection of the flexible beam system

$$4\beta - \alpha \geq 0 \tag{7.54}$$

$$8\beta - \delta_2|4\beta - \alpha| \geq 0 \tag{7.55}$$

$$4\beta - \alpha - \frac{|4\beta - \alpha|}{\delta_2} \geq 0 \tag{7.56}$$

It is not difficult to chose such α and β meeting the above inequalities. Then, by choosing $\alpha = 4$, $\beta = 1$ and $\gamma = 1$, we have

$$\lambda_3 = \min\left\{\frac{\alpha}{\beta k_2}, \frac{2k - \beta EI}{M_s}, \frac{3\alpha}{\beta k_2}, \gamma\right\} = 1 > 0 \tag{7.57}$$

The deflection of the flexible beam system with boundary control under the disturbance is shown in Fig. 7.4. From the figure, we can obtain that the vibration of the beam system is suppressed greatly within 2 s and eventually remains in the small neighborhood of its equilibrium position. Compared with the simulation results by the PD control shown in Fig. 7.3, we can see that the figure is more smooth than that of the PD control.

The end-point deflection of the flexible beam is presented in Fig. 7.5. From Fig. 7.5, it can be observed that both the PD control and the proposed control can regulate the beam's vibration. However, obviously, the response of the proposed control is faster

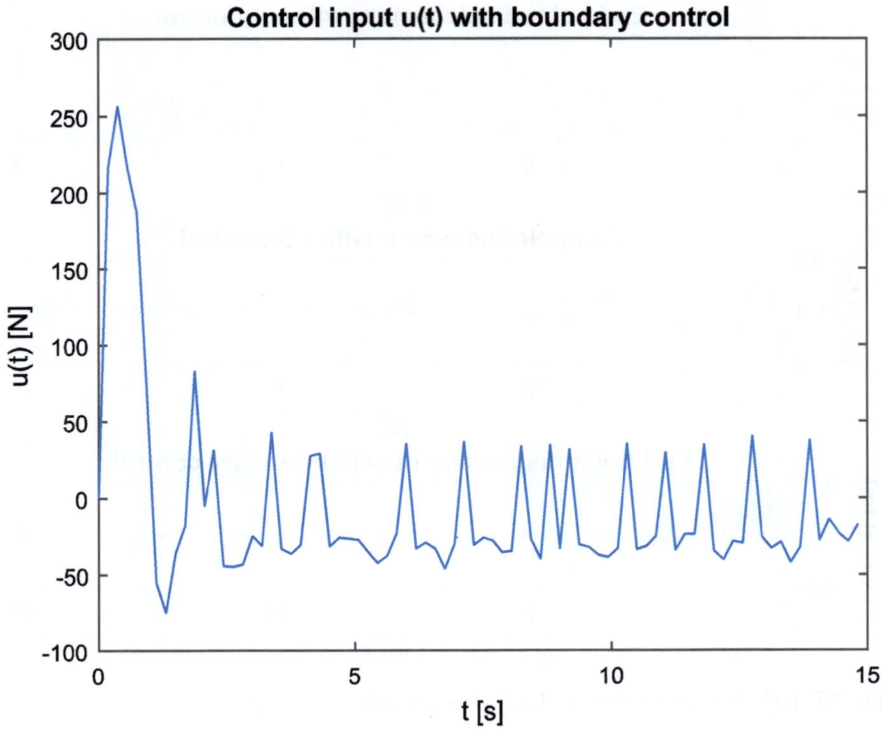

Fig. 7.6 Desired controller of the flexible beam system

than that of the PD control. From the comparison, we can conclude that we obtain a better control performance by using the proposed control law.

In addition, the curve of the designed control law $v(t)$ is obtained in Fig. 7.6.

According to the presented above results, we have that the vibration of the flexible beam system with input backlash can be reduced more effectively with the proposed boundary control law and disturbance observer. The transverse displacement $w(x, t)$ can be regulated to converge to a small neighborhood of their equilibrium positions in the presence of the external disturbance.

7.5 Conclusion

This chapter has discussed a problem of a flexible beam system with input backlash characteristic. By transforming the nonlinear input backlash to linear input and based on the linear input, the boundary control law and disturbance observer have been designed at the top of flexible beam via constructing a Lyapunov candidate function to reduce the vibration of the system and to analyze the convergence of the solution of this closed-loop system. The transverse displacement of the closed-loop system has been proved to converge to a small neighborhood of zero. By choosing appropriate parameters, numerical simulations have been presented to illustrate the effectiveness of the proposed boundary control.

References

1. C.-Y. Su, Y. Stepanenko, J. Svoboda, T.-P. Leung, Robust adaptive control of a class of nonlinear systems with unknown backlash-like hysteresis. IEEE Trans. Autom. Control **45**(12), 2427–2432 (2000)
2. M.L. Corradini, G. Orlando, Robust stabilization of nonlinear uncertain plants with backlash or dead zone in the actuator. IEEE Trans. Control Syst. Technol. **10**(1), 158–166 (2002)
3. G. Tao, P.V. Kokotovic, Adaptive control of plants with unknown hystereses. IEEE Trans. Autom. Control **40**(2), 200–212 (1995)
4. K.-H. Han, G.-O. Koh, J.-M. Sung, B.-S. Kim, Adaptive control approach for improving control systems with unknown backlash, in *2011 11th International Conference on Control, Automation and Systems* (2011), pp. 1919–1923
5. N.J. Ahmad, F. Khorrami, Adaptive control of systems with backlash hysteresis at the input, in *Proceedings of the 1999 American Control Conference*, vol. 5 (1999), pp. 3018–3022
6. J. Zhou, C. Wen, Y. Zhang, Adaptive backstepping control of a class of uncertain nonlinear systems with unknown backlash-like hysteresis. IEEE Trans. Autom. Control **49**(10), 1751–1757 (2009)
7. J. Zhou, C. Wen, *Adaptive Backstepping Control of Uncertain Systems: Nonsmooth Nonlinearities, Interactions or Time-Variations* (Springer, Berlin, 2008)
8. S.S. Ge, S. Zhang, W. He, Vibration control of an Euler-Bernoulli beam under unknown spatiotemporally varying disturbance. Int. J. Control **84**(5), 947–960 (2011)
9. G.D. Smith, *Numerical Solution of Partial Differential Equations: Finite Difference Methods* (Oxford University Press, Oxford, 1985)

Chapter 8
Distributed Control of a Flexible Beam

8.1 Introduction

Since the 1980s and 1990s, plenty of control methods and techniques have been introduced to deal with vibration problem of the flexible mechanical systems on several papers and books. In [1], by introducing feedback variables through a regulator, a state-feedback controller that contains a dynamic compensator is designed to stabilize the vibration of the flexible system. In [2], various control schemes for a single flexible robotic arm are considered and the improved control performance can be obtained by using a linear optimal controller, which verifies the efficiency of LQR optimal method. According to [3], the authors investigate the problem of synthesizing PID controllers for robust performance for a given single-input-single-output system. In [4–6], the novel active disturbance rejection control (ADRC) approach is adopted to deal with the boundary stabilization for wave equations and Euler–Bernoulli beam equations. The Riesz basis property of serially connected Timoshenko beams with joint and boundary feedback controls is investigated in [7]. However, most of the previous works on modeling of flexible mechanical systems are based on a set of finite dimensional equations, i.e., ordinary differential equations (ODEs) [8–10]. Flexible mechanical systems modeled by ODEs will neglect some system dynamics since the state of these systems depends on more than one independent variable [11]. We hence process our stability analysis through the systems described by a hybrid dynamic model containing both partial differential equations (PDEs) and ODEs [12–14]. It is more difficult for control design based on the infinite dimensionality of the system. Consequently, many approaches for PDE systems are proposed based on truncated finite dimensional modes of the system by using the finite element method [15, 16], the Galerkin's method [17, 18], the Karhunen-Loève decomposition [19–22] or the assumed modes method [23, 24]. Nonetheless, the spillover problem will happen and cause the system instability due to the ignorance of high frequency modes when dimensions are truncated [25, 26]. In order to overcome the shortcomings of spillover effect, several control techniques based on the original PDE model have been developed [14, 27–29].

W. He and J. Liu, *Active Vibration Control and Stability Analysis of Flexible Beam Systems*, https://doi.org/10.1007/978-981-10-7539-1_8

Two different boundary control methods are designed based on the flexible string in [30] and the Euler–Bernoulli beam in [31] for eliminating spillover problem, whereas, the system uniform boundedness cannot tend to zero but a close neighborhood of zero as the existing of a small constant ε or ε_0. In this chapter, we propose a distributed control based on a class of similar infinite dimensional mechanical systems without dimensional truncation, which eliminates the spillover effects and guarantees the system stability. Additionally, compared with existing boundary control techniques in [30–32], distributed control can generate high-efficient overall stability and ideal uniform ultimate boundedness. Nowadays, many types of research in regards to distributed control have been developed. In [33], a distributed parameter-dependent controller is proposed for interconnected systems with time and spatially varying parameters. In [34], the authors provide a practical distributed controller architecture from optimal controllers being obtained by solving a parameterized family of standard finite-dimensional problems and owning inherent degree of decentralization. In [35], analysis, synthesis, and implementation of distributed controllers for spatially interconnected systems, are presented. However, these distributed control methods are hardly applied to the systems with global constraint due to the ignorance of setting the limitation of boundedness.

In practical applications and industrial environment, constraint problems are increasingly common, especially in safety specifications, input saturation and mechanical stoppages. Violation of the constraints may lead to performance degradation and instability, or even cause hazards and system damage [36–38]. Therefore, the effective capability of control method to handle constraints and violation becomes more significant and crucial. To deal with the constraint issues, a number of control techniques have been developed in this field. In [39], in order to achieve high performance for the constrained tip regulation of a linear-motor-driven flexible beam, a two-loop performance-oriented tip-tracking control method is proposed, in which a feedback control law and an online trajectory replanning unit are designed for inner and outer loop, respectively. In [40, 41], the authors develop a novel method for the analysis and control of parabolic PDEs systems with input and state constraints. Galerkin's method and modal decomposition techniques are used for the derivation of ODE system that capture the dominant dynamics of the PDE system. By employing Lyapunov function and MPC formulations, the authors put forward the corresponding control laws to guarantee the closed-loop stability in the presence of constraints. For a grasping problem mentioned in [42], a boundary cooperative controller based on a hybrid model that subjects to geometric constraints is applied to obtain effective results on control of the flexible system. Furthermore, a novel concept, i.e., Barrier Lyapunov Function (BLF), has been put into use for eliminating the violation of constraint for ODE dynamics [36, 43, 44]. In [45], a bounded BLF is introduced to constrain the unknown functions within a compact superset such that the neural network approximation conditions hold. In [46], an IBLF-based controller is proposed to prevent the flexible crane system, which is represented as PDEs, from breaking the boundary output constraint. In order to suppress the risers vibration with a top tension constraint, the authors present a boundary controller for a flexible marine riser by using Integral-Barrier Lyapunov Function. Nonetheless, when confronting the more complicated global constraint, we have to come up with a new method rather than just the one handling boundary constraint.

Thus, in this chapter, we propose an Integral-Barrier Lyapunov Function to design controllers for a class of infinite dimensional systems, i.e., PDE dynamics, with global constraint. The distributed control well suppresses the unknown disturbance acting along the Euler–Bernoulli beam. Associated with this control techniques, bounded Integral-Barrier Lyapunov Function guarantees that the constraint is not transgressed and the tracking error will converge to zero when time tends to infinity. This chapter offers a great advancement in extending the Barrier Lyapunov Function theory to the infinite dimensional systems via the distributed control method.

The rest of the chapter is organized as follows. In Sect. 8.2, some notions and lemmas are introduced, some assumptions are presented, and some definitions are given. The distributed control methodologies are proposed in Sect. 8.3. Moreover, in this section, the exponential stability analysis is given and the uniform ultimate boundedness of the two closed-loop systems can be obtained via Integral-Barrier Lyapunov Function with constraint. Simulations are presented to illustrate performance of the proposed control in Sect. 8.4, where the simulation results show that the reference signal can be tracked by the movement of flexible mechanical systems and the tracking errors will tend to zero when time goes to infinity. The conclusion of this chapter is provided in Sect. 8.5.

8.2 Problem Formulation

For the convenience of stability analysis, we present the following definitions and assumptions for the subsequent development.

Definition 1 ([47]) Suppose $f\colon \mathbb{R}^n \to \mathbb{R}$ is Lebesgue measurable. We let

$$\|f\|_p = \left(\int |f(x)|^p dx\right)^{1/p}, \quad \text{if} \quad p < \infty. \tag{8.1}$$

Definition 2 ([43]) Barrier Lyapunov Function (BLF) is a scalar function $V(x)$ defined with respect to the system $\dot{x} = f(x)$ on an open region $\mathcal{D}$ containing the origin, that is continuous, positive definite. For continuous first-order partial derivatives at every point of $\mathcal{D}$, the property $V(x) \to \infty$ as x approaches the boundary of $\mathcal{D}$ holds, and $V(x(t)) \leq b$, $\forall t \geq 0$ is satisfied along the solutions of $\dot{x} = f(x)$ and some positive constant b.

Assumption 8.1 For the unknown spatiotemporally varying distributed load $f(x,t)$, we assume that there exist constants $f_1, f_2 \in \mathbb{R}^+$, such that $\|f(x,t)\|_2 \leq f_1$, and $\|\dot{f}(x,t)\|_2 \leq f_2$, $\forall t \in [0, \infty)$.

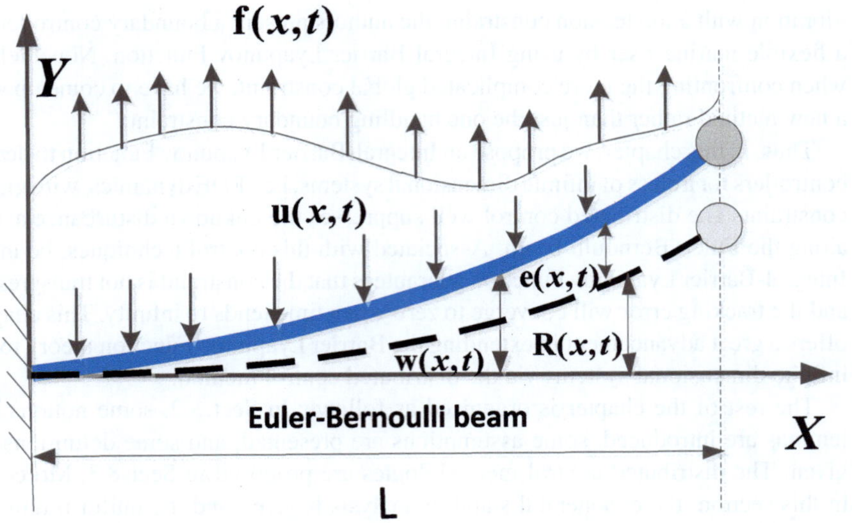

Fig. 8.1 Euler–Bernoulli system with the variable reference signal

8.3 Control Design

The control objective is to build a distributed control $u(x,t)$ to ensure that system state $w(x,t)$ can track the variable reference signal $R(x,t)$ without violation of desired constraint. The diverse Integral-Barrier Lyapunov Functions will be designed separately for analyzing the stability of two different mechanical systems. The distributed controllers will be proposed to counteract the infringement of global constraints C and guarantee the success of state tracking.

As Fig. 8.1 shows, let $w(x,t)$, $R(x,t)$, $e(x,t)$ be the transverse displacement, the desired reference, the error displacement for the Euler–Bernoulli beam. $f(x,t)$ and $u(x,t)$ are the distributed disturbances and distributed control respectively.

Consider the following Euler–Bernoulli system [31] with governing equation as

$$-c\dot{w}(x,t)+\rho\ddot{w}(x,t)+EIw''''(x,t)=Tw''(x,t)+u(x,t)+f(x,t) \quad (8.2)$$

$\forall(x,t)\in(0,L)\times[0,t_f]$, and boundary conditions as

$$w'(0,t)=0 \quad (8.3)$$

$$w''(L,t)=0 \quad (8.4)$$

$$w(0,t)=0 \quad (8.5)$$

$$-EIw'''(L,t)+Tw'(L,t)=-M_s\ddot{w}(L,t) \quad (8.6)$$

$\forall t\in[0,t_f]$.

Assumption 8.2 We assume that the desired reference $R(x, t)$ satisfies the boundary conditions, i.e.,

$$R'(0, t) = 0 \tag{8.7}$$
$$R''(L, t) = 0 \tag{8.8}$$
$$R(0, t) = 0 \tag{8.9}$$
$$-EIR'''(L, t) + TR'(L, t) = -M_s\ddot{R}(L, t) \tag{8.10}$$

$\forall t \in [0, t_f]$.

Define $e(x, t) = w(x, t) - R(x, t)$. Differentiating the tracking error twice by time and using the governing equation (8.2), we obtain the error dynamics as

$$\begin{aligned}\rho\ddot{e}(x, t) + c\dot{e}(x, t) =& Te''(x, t) - EIe''''(x, t) + u(x, t) + f(x, t) - \rho\ddot{R}(x, t)\\ &- EIR''''(x, t) + TR''(x, t) - c\dot{R}(x, t)\end{aligned} \tag{8.11}$$

$\forall (x, t) \in (0, L) \times [0, t_f]$, where c is a positive constant. The boundary conditions are given as

$$e'(0, t) = 0 \tag{8.12}$$
$$e''(L, t) = 0 \tag{8.13}$$
$$e(0, t) = 0 \tag{8.14}$$
$$-EIe'''(L, t) + Te'(L, t) = -d_s\dot{e}(L, t) - M_s\ddot{e}(L, t) \tag{8.15}$$

$\forall t \in [0, t_f]$, where d_s is a positive constant. Then we consider the following Lyapunov function candidate

$$\begin{aligned}V(t) =& \frac{1}{2}\rho\int_0^L [P(x, t)]^2 \ln\frac{A^2}{A^2 - e^2(x, t)}dx + \frac{1}{2}\int_0^L [e(x, t)]^2 \ln\frac{A^2}{A^2 - e^2(x, t)}dx\\ &+ \frac{1}{2}\int_0^L [e(x, t)]^2 dx\end{aligned} \tag{8.16}$$

where an auxiliary term $P(x, t)$ is defined as

$$P(x, t) = e(x, t) + \dot{e}(x, t) \tag{8.17}$$

Remark 8.1 We set that the reference signal $|R(x, t)| \leq N_0$, where N_0 is a positive constant and $N_0 < C$. As long as the tracking error $e(x, t)$ can be limited within a positive constant $A = C - N_0$, we can conclude that $|w(x, t)| \leq |e(x, t)| + |R(x, t)| < A + N_0 = C$, i.e., $w(x, t)$ is constrained within C.

The controller is designed as

$$\begin{aligned} u(x,t) =& -(\rho-c)\dot{e}(x,t) - Te''(x,t) + EIe''''(x,t) - k_1P(x,t) - f_1 \\ &+ \rho\ddot{R}(x,t) + EIR''''(x,t) - TR''(x,t) + c\dot{R}(x,t) \\ &- \left[\rho P(x,t)\frac{e(x,t)\dot{e}(x,t)}{A^2-e^2(x,t)}\Big/\ln\frac{A^2}{A^2-e^2(x,t)}\right] \\ &- e(x,t) - \left[\frac{1}{2}P(x,t)\frac{e^2(x,t)}{A^2-e^2(x,t)}\Big/\ln\frac{A^2}{A^2-e^2(x,t)}\right] \\ &- e(x,t)\Big/\ln\frac{A^2}{A^2-e^2(x,t)} \end{aligned} \tag{8.18}$$

where sgn(·) denotes the signum function, k_1 is a positive control gain.

Differentiating the Lyapunov function candidate by time and substituting the above controller, we can obtain

$$\begin{aligned} \dot{V}(t) =& -k_1\int_0^L [P(x,t)]^2 \ln\frac{A^2}{A^2-e^2(x,t)}dx - \int_0^L [e(x,t)]^2 \ln\frac{A^2}{A^2-e^2(x,t)}dx \\ &- \frac{1}{2}\int_0^L [e(x,t)]^2\frac{e^2(x,t)+\dot{e}^2(x,t)}{A^2-e^2(x,t)}dx - \int_0^L [e(x,t)]^2 dx \\ \leq& -\lambda V(t) \end{aligned} \tag{8.19}$$

where

$$\lambda = \min\left\{\frac{2k_1}{\rho}, 2\right\} > 0 \tag{8.20}$$

Theorem 8.1 *For the Euler–Bernoulli beam system described by governing equation* (8.2) *and boundary conditions* (8.3)–(8.6), *with the desired reference signal* $R(x,t)$ *and proposed distributed control* (8.18), *under bounded initial condition, if the control parameters are selected appropriately, the tracking error* $e(x,t)$ *will exponentially stable.*

Proof Multiplying (8.19) by $e^{\lambda t}$ yields

$$\frac{d}{dt}(Ve^{\lambda t}) \leq 0 \tag{8.21}$$

Integrating of the above inequality, we obtain

$$V \leq V(0)e^{-\lambda t} \in \mathcal{L}_\infty \tag{8.22}$$

which implies V is bounded. According to Definition 1, we have

$$\int_0^L [e(x,t)]^2 dx = ||e(x,t)||_2^2 \le 2V(t) \le 2V(0)e^{\lambda t} \tag{8.23}$$

Furthermore, we obtain

$$||e(x,t)||_2 \le \sqrt{2V(0)e^{\lambda t}}, \forall x \in [0, L] \tag{8.24}$$

and then

$$\lim_{t \to \infty} ||e(x,t)||_2 = 0 \tag{8.25}$$

8.4 Numerical Simulations

In this section, the Euler–Bernoulli system is simulated to reveal the effectiveness of the proposed distributed controller respectively. To demonstrate the effectiveness of the proposed approach for a class of flexible mechanical systems with the global constraint, we choose finite difference (FD) method to conduct the numerical simulation since FD method can be applied to obtain the approximate solution of the Euler–Bernoulli beam system, when there is no analytical solution. The spatial step size and temporal step size are set as $\Delta x = 0.03$ m and $\Delta t = 2.5 \times 10^{-4}$ s. After the approximation process, we can obtain the results of the corresponding simulation.

In this case, the parameters of the flexible Euler–Bernoulli beam system are given in the Table 8.1.

The external disturbance $f(x,t)$ and initial conditions are described as

$$f(x,t) = [15 + \sin(\pi x t) + \sin(2\pi x t) + \sin(3\pi x t)]\frac{x}{L} \tag{8.26}$$

$$w(x,0) = 0 \tag{8.27}$$

$$\dot{w}(x,0) = 0 \tag{8.28}$$

Table 8.1 Parameters of the Euler–Bernoulli beam

Parameter	Description	Value
L	Length of the beam	1 m
M_s	Mass of the tip payload	5 kg
T	Tension of the beam	6 N
EI	Bending stiffness of the beam	5 Nm^2
ρ	Mass per unit length of the beam	1 kg/m

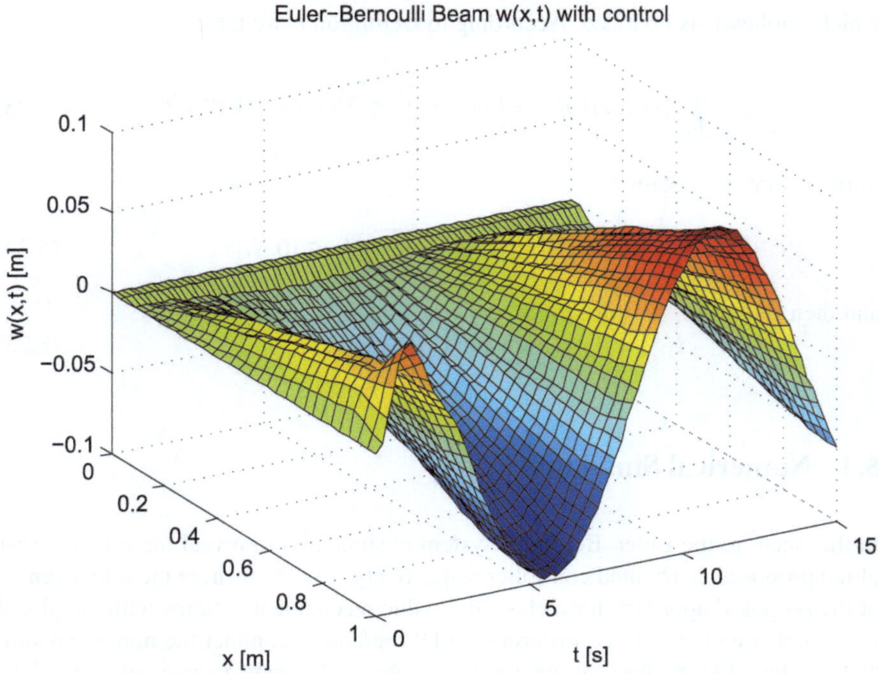

Fig. 8.2 Displacement of the flexible beam system with control

The parameters are given as $c = 1$, $k_1 = 2000$, $C = 0.3$, $d_s = 1$ and $A = 0.2$. The desired reference signal $R(x, t)$ is set as below and shown in Fig. 8.3.

$$\rho \ddot{R}(x, t) = T_0 R''(x, t) + c_0 \dot{R}(x, t) \tag{8.29}$$

where $T_0 = 10$, $c_0 = 0$, and the corresponding initial conditions are

$$R(x, 0) = 0.1x \tag{8.30}$$

$$\dot{R}(x, 0) = 0 \tag{8.31}$$

The reference signal is trackable at the position $x = L$ within 3s in Fig. 8.5. Accordingly, the tracking error shown in Fig. 8.4 can be regulated and remained in a small range around zero. And the proposed control law can counteract the infringement of the global constraint C which is displayed in Fig. 8.7. For clarity, in Fig. 8.6, the tracking error can tend to zero within 4 seconds and the error constraint A is not violated. Comparing Fig. 8.2 with Fig. 8.3, it is easy to conclude that the deflection $w(x, t)$ can track the reference $R(x, t)$ very well without violation of the desired constraint.

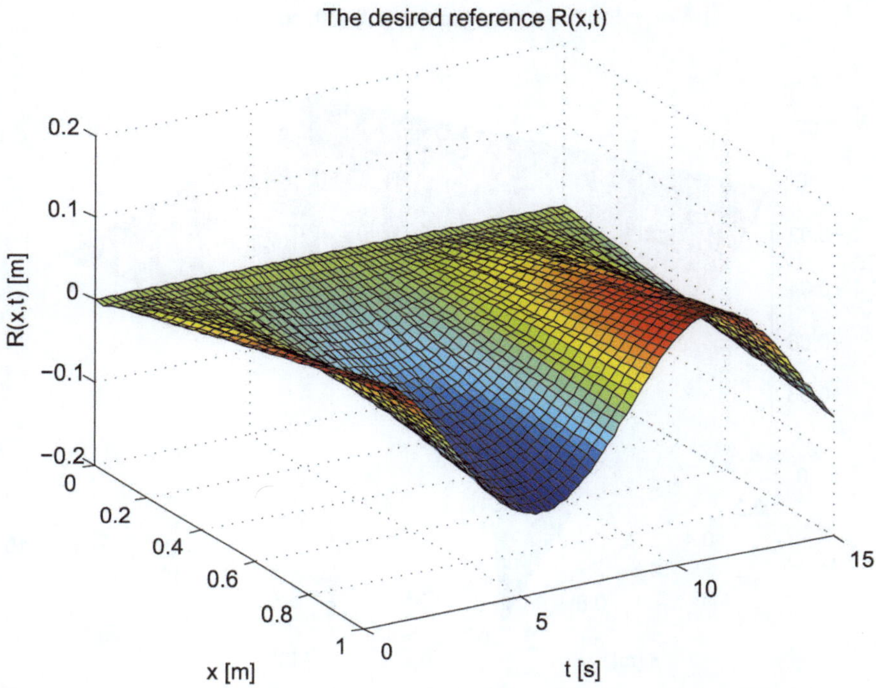

Fig. 8.3 The desired reference signal of flexible beam system

8.5 Conclusion

In this chapter, we have presented the control design for a flexible Euler–Bernoulli beam system with global constraint and uncertain tip payload. To prevent the constraint violation, a novel Integral-Barrier Lyapunov Function is employed for the control design and stability analysis. We have proposed a distributed control so that the deflection of the mechanical systems can track a desired reference signal. Exponential stability has been well achieved without violation of the constraint. Extensive simulations have been provided to illustrate the performance of the control system. This chapter has represented an important step in extending the Barrier Lyapunov Function theory to the infinite dimensional system.

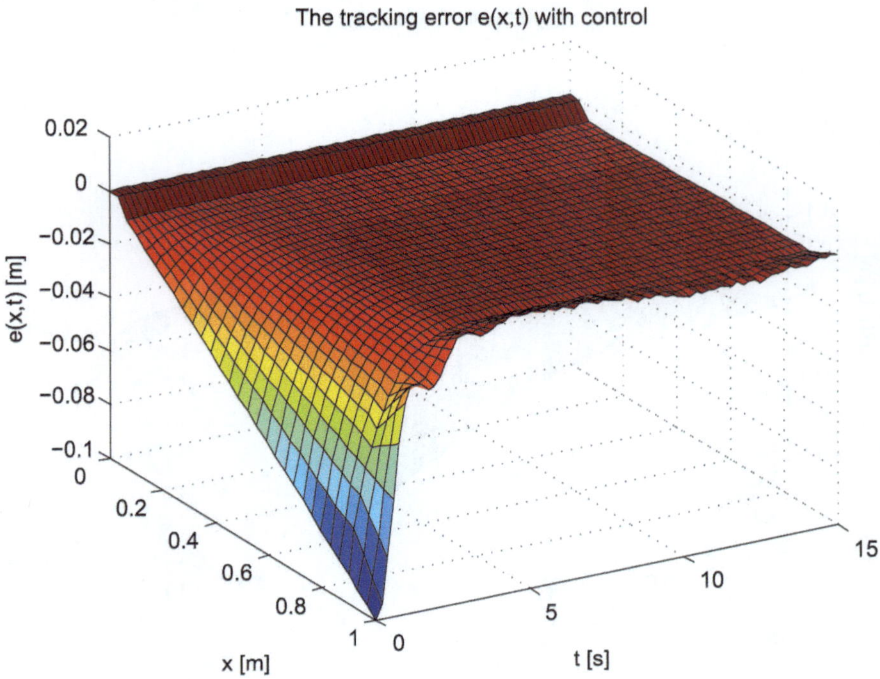

Fig. 8.4 The error of the flexible beam system with control

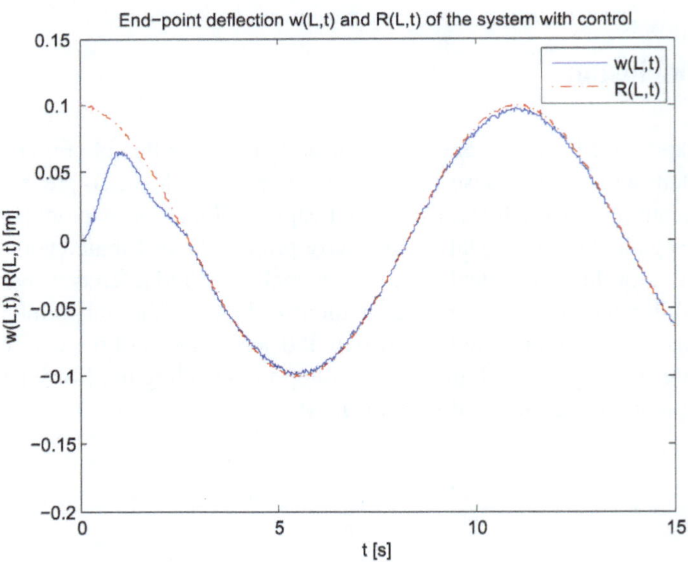

Fig. 8.5 End-point deflection and desired reference of the flexible beam system with control

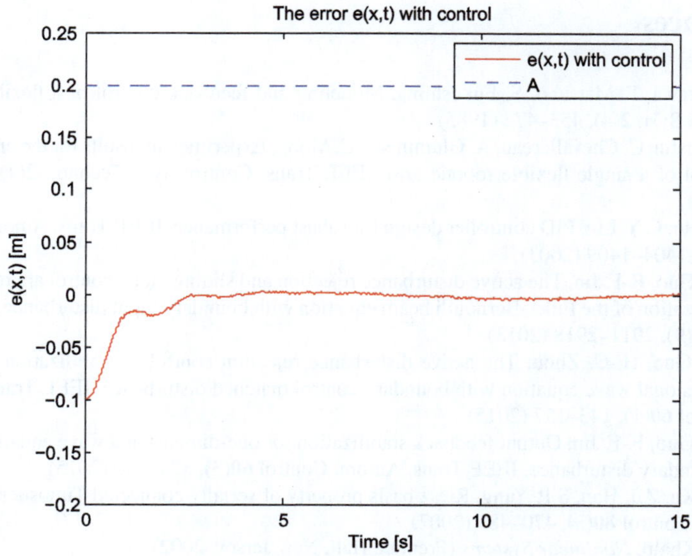

Fig. 8.6 The error and error constraint of the flexible beam system with control

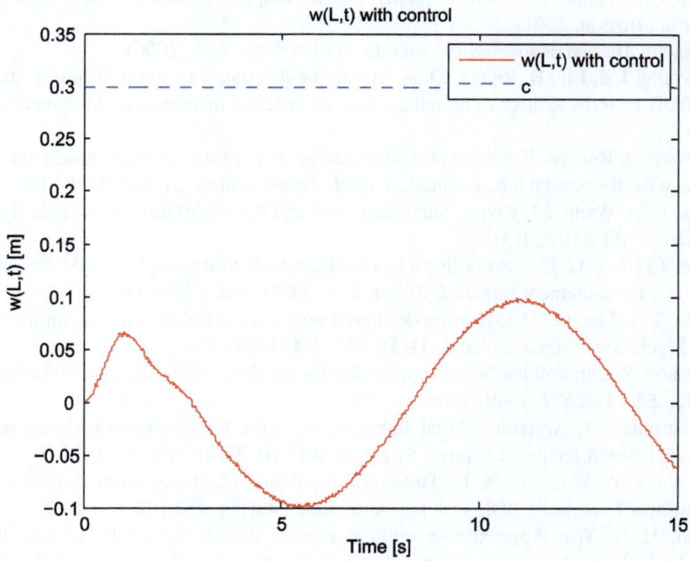

Fig. 8.7 End-point deflection and global constraint of the flexible beam system with control

References

1. Y. Sakawa, F. Matsuno, S. Fukushima, Modeling and feedback control of a flexible arm. J. Robot. Syst. **2**(4), 453–472 (1985)
2. Y. Aoustin, C. Chevallereau, A. Glumineau, C. Moog, Experimental results for the end-effector control of a single flexible robotic arm. IEEE Trans. Control Syst. Technol. **2**(4), 371–381 (1994)
3. M.T. Ho, C.Y. Lin, PID controller design for robust performance. IEEE Trans. Autom. Control **48**(8), 1404–1409 (2003)
4. B.-Z. Guo, F.-F. Jin, The active disturbance rejection and sliding mode control approach to the stabilization of the Euler-Bernoulli beam equation with boundary input disturbance. Automatica **49**(9), 2911–2918 (2013)
5. B.-Z. Guo, H.-C. Zhou, The active disturbance rejection control to stabilization for multi-dimensional wave equation with boundary control matched disturbance. IEEE Trans. Autom. Control **60**(1), 143–157 (2015)
6. B.-Z. Guo, F.-F. Jin, Output feedback stabilization for one-dimensional wave equation subject to boundary disturbance. IEEE Trans. Autom. Control **60**(3), 824–830 (2015)
7. G.Q. Xu, Z.J. Han, S.P. Yung, Riesz basis property of serially connected Timoshenko beams. Int. J. Control **80**(3), 470–485 (2007)
8. H.K. Khalil, *Nonlinear Systems* (Prentice Hall, New Jersey, 2002)
9. M. Krstic, I. Kanellakopoulos, P. Kokotovic, *Nonlinear and Adaptive Control Design* (Wiley, New York, 1995)
10. S.S. Ge, C.C. Hang, T.H. Lee, T. Zhang, *Stable Adaptive Neural Network Control* (Kluwer Academic, Boston, 2001)
11. J.D. Logan, *Applied Mathematics*, 3rd edn. (Wiley, New York, 2006)
12. J.-M. Wang, J.-J. Liu, B. Ren, J. Chen, Sliding mode control to stabilization of cascaded heat PDE-ODE systems subject to boundary control matched disturbance. Automatica **52**, 23–34 (2015)
13. J.M. Wang, B. Ren, M. Krstic, Stabilization and gevrey regularity of a Schrödinger equation in boundary feedback with a heat equation. IEEE Trans. Autom. Control **57**(1), 179–185 (2012)
14. B. Ren, J.-M. Wang, M. Krstic, Stabilization of an ODE-Schrödinger cascade. Syst. Control Lett. **62**(6), 503–510 (2013)
15. S.S. Ge, T.H. Lee, G. Zhu, A nonlinear feedback controller for a single-link flexible manipulator based on a finite element model. J. Robot. Syst. **14**(3), 165–178 (1997)
16. S.S. Ge, T.H. Lee, G. Zhu, Non-model-based position control of a planar multi-link flexible robot. Mech. Syst. Signal Process. **11**(5), 707–724 (1997)
17. A. Armaou, P. Christofides, Wave suppression by nonlinear finite-dimensional control. Chem. Eng. Sci. **55**(14), 2627–2640 (2000)
18. P. Christofides, A. Armaou, Global stabilization of the Kuramoto-Sivashinsky equation via distributed output feedback control. Syst. Control Lett. **39**(4), 283–294 (2000)
19. H.-N. Wu, J.-W. Wang, H.-X. Li, Design of distributed H_∞ fuzzy controllers with constraint for nonlinear hyperbolic PDE systems. Automatica **48**(10), 2535–2543 (2012)
20. B. Luo, H.-N. Wu, Approximate optimal control design for nonlinear one-dimensional parabolic PDE systems using empirical eigenfunctions and neural network. IEEE Trans. Syst. Man Cybern. Part B: Cybern. **42**(6), 1538–1549 (2012)
21. J. Wang, H.-N. Wu, H.-X. Li, Distributed proportional spatial derivative control of nonlinear parabolic systems via fuzzy PDE modeling approach. IEEE Trans. Syst. Man Cybern. Part B: Cybern. **42**(3), 927–938 (2012)
22. H.-N. Wu, H.-X. Li, Finite-dimensional constrained fuzzy control for a class of nonlinear distributed process systems. IEEE Trans. Syst. Man Cybern. Part B (Cybern.) **37**(5), 1422–1430 (2007)
23. B. Bhikkaji, S.O.R. Moheimani, I. Petersen, A negative imaginary approach to modeling and control of a collocated structure. IEEE/ASME Trans. Mechatron. **17**(4), 717–727 (2012)

24. E. Pereira, S.S. Aphale, V. Feliu, S.R. Moheimani, Integral resonant control for vibration damping and precise tip-positioning of a single-link flexible manipulator. IEEE/ASME Trans. Mechatron. **16**(2), 232–240 (2011)
25. M.J. Balas, Active control of flexible systems. J. Optim. Theory Appl. **25**, 415–436 (1978)
26. L. Meirovitch, H. Baruh, On the problem of observation spillover in self-adjoint distributed systems. J. Optim. Theory Appl. **39**(2), 269–291 (1983)
27. M. Krstic, A. Smyshlyaev, *Boundary Control of PDEs: A Course on Backstepping Designs* (Society for Industrial and Applied Mathematics, Philadelphia, 2008)
28. D. Huang, J.-X. Xu, X. Li, C. Xu, M. Yu, D-type anticipatory iterative learning control for a class of inhomogeneous heat equations. Automatica **49**(8), 2397–2408 (2013)
29. W. He, S.S. Ge, B.V.E. How, Y.S. Choo, *Dynamics and Control of Mechanical Systems in Offshore Engineering* (Springer, London, 2014)
30. W. He, S.S. Ge, Robust adaptive boundary control of a vibrating string under unknown time-varying disturbance. IEEE Trans. Control Syst. Technol. **20**(1), 48–58 (2012)
31. S.S. Ge, S. Zhang, W. He, Vibration control of an Euler-Bernoulli beam under unknown spatiotemporally varying disturbance. Int. J. Control **84**(5), 947–960 (2011)
32. G.Q. Xu, H. Wang, Stabilisation of Timoshenko beam system with delay in the boundary control. Int. J. Control **86**(6), 1165–1178 (2013)
33. F. Wu, Distributed control for interconnected linear parameter-dependent systems. IEE Proc. Control Theory Appl. **150**(5), 518–527 (2003)
34. B. Bamieh, F. Paganini, M.A. Dahleh, Distributed control of spatially invariant systems. IEEE Trans. Autom. Control **47**(7), 1091–1107 (2002)
35. R. D'Andrea, G.E. Dullerud, Distributed control design for spatially interconnected systems. IEEE Trans. Autom. Control **48**(9), 1478–1495 (2003)
36. K.P. Tee, B. Ren, S.S. Ge, Control of nonlinear systems with time-varying output constraints. Automatica **47**(11), 2511–2516 (2011)
37. W. He, S.S. Ge, Vibration control of a flexible beam with output constraint. IEEE Trans. Ind. Electron. **62**(8), 5023–5030 (2015)
38. W. He, Y. Chen, Z. Yin, Adaptive neural network control of an uncertain robot with full-state constraints. IEEE Trans. Cybern. **46**(3), 620–629 (2016)
39. L. Lu, Z. Chen, B. Yao, Q. Wang, A two-loop performance oriented tip tracking control of a linear motor driven flexible beam system with experiments. IEEE Trans. Ind. Electron. **60**(3), 1011–1022 (2013)
40. N.H. El Farra, A. Armaou, P.D. Christofides, Analysis and control of parabolic PDE systems with input constraints. Automatica **39**(4), 715–725 (2003)
41. S. Dubljevic, N.H. El Farra, P. Mhaskar, P.D. Christofides, Predictive control of parabolic PDEs with state and control constraints. Int. J. Robust Nonlinear Control **16**(16), 749–772 (2006)
42. K. Endo, F. Matsuno, H. Kawasaki, Simple boundary cooperative control of two one-link flexible arms for grasping. IEEE Trans. Autom. Control **54**(10), 2470–2476 (2009)
43. K.P. Tee, S.S. Ge, E. Tay, Barrier Lyapunov functions for the control of output-constrained nonlinear systems. Automatica **45**(4), 918–927 (2009)
44. W. He, C. Sun, S.S. Ge, Top tension control of a flexible marine riser by using integral-barrier Lyapunov function. IEEE/ASME Trans. Mechatron. **2**(20), 497–505 (2015)
45. B. Ren, S.S. Ge, K.P. Tee, T. Lee, Adaptive neural control for output feedback nonlinear systems using a barrier Lyapunov function. IEEE Trans. Neural Netw. **21**(8), 1339–1345 (2010)
46. W. He, S. Zhang, S.S. Ge, Adaptive control of a flexible crane system with the boundary output constraint. IEEE Trans. Ind. Electron. **61**(8), 4126–4133 (2014)
47. E. Hewitt, K. Stromberg, *Real and Abstract Analysis: A Modern Treatment of the Theory of Functions of a Real Variable* (Springer, Berlin, 1929)

Chapter 9
Iterative Learning Control of a Flexible Beam

9.1 Introduction

Iterative learning control (ILC) is widely used in engineering, especially in the repetitive processes, including robot arm manipulators, chemical batch processes and reliability testing rigs [1, 2]. By learning from the past control experience, system performances are improved [3]. ILC is employed without perfect knowledge of the target system, and possesses relative simple structure and almost model-free nature [4, 5]. Most ILC schemes are proposed to track periodic trajectories and reject periodic disturbances under the identical initial condition [6].

Based on the contraction mapping method, the ILC, including P-type [7], D-type [8], PD-type [9], PID-type [10], etc., is used for the system with the global Lipschitz continuity. In order to break such limitations, composite energy function (CEF) is used to propose ILC schemes for the target system with the local Lipschitz continuity. The CEF is relative to the system energy and thus can be used to prove the stability of the closed-loop system. Meanwhile, alignment condition is utilized, where the initial states at the current iteration are aligned with the final states at the previous iteration. Frequently confronted with the uncertainties of the considered system, it is thus indispensable to incorporate adaptive control into the ILC methodology. In [11], a discrete-time adaptive ILC law is designed to tackle time-varying parametric uncertainties and guarantee the asymptotic learning convergence. In [12], by using an adaptive robust ILC scheme, the structured system uncertainties are addressed in the presence of input dead-zone. Rigid robot manipulators perform repetitive tasks in [13], where adaptive ILC schemes are proposed to track the desired trajectories and deal with the unknown parameters and disturbances. In [14], adaptive ILC is proposed for an industrial robot system under the measurement disturbance. In [15], a unified framework of adaptive ILC is designed for uncertain nonlinear systems under alignment condition. However, all the above systems are the ordinary differential equation (ODE) systems. Although a class of PDE systems are considered in [16–18], ILC laws are proposed to merely achieve the learning convergence for the target

W. He and J. Liu, *Active Vibration Control and Stability Analysis of Flexible Beam Systems*, https://doi.org/10.1007/978-981-10-7539-1_9

systems without input saturation and aperiodic disturbances, particularly without aperiodic spatiotemporally varying disturbances.

Based on Lyapunov's method [19–21], adaptive boundary control has been utilized to address the vibration suppression [22, 23] for flexible string [24, 25], Euler-Bernoulli beam [26, 27], Timoshenko beam [28] and robot arm system [29]. However, with the distributed disturbance considered, adaptive boundary control cannot guarantee the exponential stability or asymptotic stability of the closed-loop system [30]. In this chapter, the control problem is addressed by an adaptive boundary iterative learning control (ABILC) law, which aims to (I) tackle the input constraint, (II) reject the aperiodic distributed disturbance and the aperiodic boundary disturbance, (III) cope with the uncertainty of the system parameter, and (IV) guarantee the asymptotic convergence of the displacements of the closed-loop system.

In practice, constraints are ubiquitous in physical systems, in the forms of input saturations [31], output constraints [32], etc., for the limitations of energy and magnitude in physical actuators, sensors and interfacing devices [33]. Violations of the constraints may degrade system performance, give rise to undesired vibrations, and even result in system paralysis [34]. Studying the system with the input constraint is therefore meaningful and indispensable. There are many works focusing on how to tackle the input constraint, for example, [35] utilizes linear matrix inequalities (LMIs) and [36] introduces nested saturation input functions. However, the LMIs are applicable only for linear systems. The nested saturation input functions always consist of symbolic functions and are not continuously differentiable, therefore, give rise to the undesired chatters. In this paper, hyperbolic tangent functions are adopted in the CEF to reduce the chatters.

The rest of the chapter is organized as follows. In Sect. 9.2, the model of the Euler-Bernoulli beam system is given under some assumptions. An ABILC scheme is designed in Sect. 9.3. Theorem 9.1 is proved in Sect. 9.4. Simulations are provided to illustrate the effectiveness of the proposed ABILC law in Sect. 9.5. The conclusion follows in Sect. 9.6.

9.2 Problem Formulation

In this section, the Euler-Bernoulli beam system is considered under Assumptions 9.1–9.3.

Let $j \in \mathbb{N}$ represents the iteration number. $t \in [0, T_b]$ is the time parameter, where T_b is a positive constant. ρ denotes the unit mass per unit length for the Euler-Bernoulli beam. T is the tension of the flexible beam. EI denotes the bending stiffness of the beam. L is the length of the beam. M expresses the mass of the tip load. $f_j(x, t)$ and $d_j(t)$ express the aperiodic distributed disturbance and the aperiodic boundary disturbance at the j th iteration, respectively. $\dot{w}_j(x, t)$ is the velocity of the beam at the j th iteration. $\ddot{w}_j(x, t)$ denotes the accelerated speed of the flexible beam. $w'_j(x, t)$ is the slope of the beam. $\dot{w}_j(L, t)$ denotes the velocity of the tip payload. $\ddot{w}_j(L, t)$ is the accelerated speed of the tip payload.

As shown in [37], the Euler-Bernoulli beam system is modeled as a distributed parameter system, which is expressed by one partial differential equation (PDE) and four boundary conditions. The Euler-Bernoulli beam system at the j th iteration is represented by

$$\rho\ddot{w}_j(x,t) + EIw_j''''(x,t) = f_j(x,t) + Tw_j''(x,t), \tag{9.1}$$

$$w_j(0,t) = 0, \tag{9.2}$$

$$w_j'(0,t) = 0, \tag{9.3}$$

$$w_j''(L,t) = 0, \tag{9.4}$$

$$M\ddot{w}_j(L,t) - EIw_j'''(L,t) = u^*\tanh\left(\frac{u_j(t)}{u^*}\right) + d_j(t) - Tw_j'(L,t), \tag{9.5}$$

for $\forall(x,t) \in [0,L) \times [0,T_b]$ and $j \in \{0,1,2,3,\dots\}$.

Assumption 9.1 ([38]) The alignment condition holds for all iterations, i.e., $w_j(x,0) = w_{j-1}(x,T_b)$ and $\dot{w}_j(x,0) = \dot{w}_{j-1}(x,T_b)$ for $\forall x \in [0,L]$ and $j \in \mathbb{N}$.

Assumption 9.2 Considering the finite energy of the boundary disturbance, $d_j(t)$ is thus bounded for $t \in [0,T_b]$ and $j \to +\infty$. An unknown positive constant $\bar{d}$ exists with $|d_j(t)| \le \bar{d}$ for $\forall t \in [0,T_b]$ and $\forall j \in \mathbb{N}$.

Assumption 9.3 Since the distributed disturbance is continuous and has the finite energy for $(x,t) \in [0,L] \times [0,T_b]$ and $j \to +\infty$, an unknown positive constant $\bar{f}$ exists with $|f_j(x,t)| \le \bar{f}$.

9.3 Control Design

In this section, input constraint is tackled through the designed control law (9.6). Three adaptive laws are proposed in the iteration domain. The BILC scheme (9.6) is designed to suppress the vibrations.

The BILC scheme is proposed as follows

$$\begin{cases} u_j(t) = & U_j(t) - \nu_1\bar{f}L\tanh(\nu_5\dot{w}_j(L,t)) - \left[\nu_2\bar{d} + \nu_3^*(1-\tanh(1))\right]^* \\ & \tanh(\nu_4\dot{w}_j(L,t)), \\ U_j(t) = & u^*\tanh\left(\frac{U_{j-1}(t)}{u^*}\right) - \gamma\dot{w}_j(L,t), \end{cases} \tag{9.6}$$

where $U_{-1}(t) = 0$, u^* is a positive constant, and ν_1–ν_5 and γ are positive constants.

Property 9.1 *For the proposed BILC law* (9.6), *the following inequality is obtained [38].*

$$|u_j(t) - u^*\mathrm{sat}\left(\frac{u_j(t)}{u^*}, 1\right)| \leq (\gamma + \nu_1 \bar{f} L \nu_5 + (\nu_2 \bar{d} + \nu_3 u^*(1 - \tanh(1)))\nu_4)|\dot{w}_j(L, t)| \tag{9.7}$$

where the saturated function is defined as

$$\mathrm{sat}\left(\frac{u_j(t)}{u^*}, 1\right) = \begin{cases} \mathrm{sgn}(u_j(t)), & \text{if } |u_j(t)| \geq u^*, \\ \frac{u_j(t)}{u^*}, & \text{if } |u_j(t)| < u^*, \end{cases} \tag{9.8}$$

From (9.8), *we can therefore obtain*

$$|u^* \tanh\left(\frac{u_j(t)}{u^*}\right) - u^*\mathrm{sat}(\frac{u_j(t)}{u^*}, 1)| \leq u^*(1 - \tanh(1)) \tag{9.9}$$

Remark 9.1 Different from input saturation in (9.8), constrained input is expressed by $u^* \tanh(\frac{u_j(t)}{u^*})$ which is differentiable and can not reach the constraint values. The input saturation can be addressed through auxiliary systems [39, 40], or hyperbolic tangent functions [36]. In this paper, input constraint in (9.6) is addressed by dividing it into a saturation input (9.8) and a bounded term, which can be regarded as another boundary disturbance with known bound.

Remark 9.2 By comparison, the main contributions are summarized as follows

(i) In the literature of ILC, serval works such as [8, 17, 18, 41] propose ILC for second-order PDE systems. In [8, 41], unknown nonlinear functions were assumed to be Lipschitz continuous. References [17, 18] did not consider external disturbances. In our previous works [42, 43], symbolic functions are used in the BILC law to tackle input constraint and reject disturbances. However, this paper proposes a unified but differentiable BILC scheme to tackle such issues, which can be described in detail as follows

   (i-a) A BILC law is designed for a typical flexible Euler-Bernoulli beam structure and the closed-loop system converges to zero along the iteration axis;
   (i-b) Hyperbolic tangent function are utilized to tackle input constraint, therefore avoiding chatters produced by symbolic functions;
   (i-c) Based on hyperbolic tangent function, spatiotemporally varying disturbances are compensated for and time-varying disturbances are rejected.

(ii) Compared with proportion-integral-derivative control, more issues, including input constraint boundary disturbances, and even spatiotemporally varying disturbances, are addressed by the BILC law smoothly;
(iii) Compared with model-based control methodologies, the BILC scheme (9.6) is mostly model-free.

9.4 Convergence Analysis

For the Euler-Bernoulli beam system, by using the proposed ABILC law (9.6), the convergence property is proved based on a time-weighted Lyapunov-Krasovskii-like CEF in the presence of the input constraint, the aperiodic distributed disturbance, the aperiodic boundary disturbance and the unknown system parameter, which is summarized in the following theorem.

Theorem 9.1 *For the Euler-Bernoulli beam system described by* (9.1)–(9.5), *under Assumptions 9.1–9.3, given that the initial system states are bounded, by using the ABILC scheme* (9.6), *the following properties hold*

(I) All the system states, including $\dot{w}_j(x,t)$, $w'_j(x,t)$ *and* $w''_j(x,t)$, *are bounded in each iteration;*

(II) The displacements of the Euler-Bernoulli beam converge to zero, namely, $\lim_{j\to+\infty}|w_j(x,t)| = 0$ *for* $\forall(x,t)\in[0,L]\times[0,T_b]$*;*

(III) The distributed disturbance and the boundary disturbance are rejected;

(IV) The input constraint is tackled by using a hyperbolic tangent function and a saturation function;

Proof Define the following Lyapunov-Krasovskii-like CEF

$$E_j(t) = V_j(t) + \frac{\alpha}{2\gamma}\int_0^t e^{-\lambda\tau}[U_j(\tau)]^2 d\tau, \tag{9.10}$$

where λ is a positive constant.

$V_j(t)$ is related to the system energy and is then expressed by

$$V_j(t) = V_{1j}(t) + V_{2j}(t), \tag{9.11}$$

where $V_{1j}(t)$ and $V_{2j}(t)$ are defined as

$$\begin{aligned} V_{1j}(t) &= \frac{\alpha\rho}{2}\int_0^L e^{-\lambda t}[\dot{w}_j(x,t)]^2 dx + \frac{\alpha EI}{2}\int_0^L e^{-\lambda t}[w''_j(x,t)]^2 dx \\ &\quad + \frac{\alpha T}{2}\int_0^L e^{-\lambda t}[w'_j(x,t)]^2 dx, \qquad (9.12) \\ V_{2j}(t) &= \frac{\alpha M}{2} e^{-\lambda t}[\dot{w}_j(L,t)]^2. \qquad (9.13) \end{aligned}$$

The proof consists of four steps, namely, the derivative of $V_j(t)$, the boundedness of $E_0(T_b)$, the boundedness of all the system states in each iteration and the convergence of $E_j(T_b)$.

Step I: Derivative of $V_j(t)$

Substituting (9.1) into $\dot{V}_{1j}(t)$ leads to

$$\dot{V}_{1j}(t) = B_{1j}(t) + B_{2j}(t) + B_{3j}(t) + B_{4j}(t), \tag{9.14}$$

where $B_{1j}(t)$–$B_{4j}(t)$ are defined as follows

$$B_{1j}(t) = -\frac{\alpha\lambda\rho}{2}\int_0^L e^{-\lambda t}[\dot{w}_j(x,t)]^2dx - \frac{\alpha\lambda EI}{2}\int_0^L e^{-\lambda t}[w_j''(x,t)]^2dx \\ -\frac{\alpha\lambda T}{2}\int_0^L e^{-\lambda t}[w_j'(x,t)]^2dx, \tag{9.15}$$

$$B_{2j}(t) = \alpha\int_0^L e^{-\lambda t}\dot{w}_j(x,t)\Big[-EIw_j''''(x,t) + Tw_j''(x,t) + f_j(x,t)\Big]dx, \tag{9.16}$$

$$B_{3j}(t) = \alpha EI\int_0^L e^{-\lambda t}w_j''(x,t)\dot{w}_j''(x,t)dx, \tag{9.17}$$

$$B_{4j}(t) = \alpha T\int_0^L e^{-\lambda t}w_j'(x,t)\dot{w}_j'(x,t)dx. \tag{9.18}$$

Integrating by parts and substituting (9.2)–(9.4), (9.17) is rewritten as

$$B_{3j}(t) = -\alpha EIe^{-\lambda t}w_j'''(L,t)\dot{w}_j(L,t) + \alpha EI\int_0^L e^{-\lambda t}w_j''''(x,t)\dot{w}_j(x,t)dx. \tag{9.19}$$

Similarly, (9.18) is reconstructed as

$$B_{4j}(t) = \alpha Te^{-\lambda t}w_j'(L,t)\dot{w}_j(L,t) - \alpha T\int_0^L e^{-\lambda t}w_j''(x,t)\dot{w}_j(x,t)dx. \tag{9.20}$$

Differentiating $V_{2j}(t)$ with respect to t and substituting (9.5), we have

$$\dot{V}_{2j}(t) = -\frac{\alpha M\lambda}{2}e^{-\lambda t}[\dot{w}_j(L,t)]^2 + \alpha Me^{-\lambda t}\dot{w}_j(L,t)\ddot{w}_j(L,t) \\ = -\frac{\alpha\lambda M}{2}e^{-\lambda t}[\dot{w}_j(L,t)]^2 + \alpha e^{-\lambda t}\dot{w}_j(L,t)[EIw_j'''(L,t) + u^*\tanh(\frac{u_j(t)}{u^*}) \\ +d_j(t) - Tw_j'(L,t)]. \tag{9.21}$$

Using (9.6), Lemma 2.2 and the Property 9.1, and combining (9.20) and (9.21), the time derivative of $V_j(t)$ is then expressed by

$$\dot{V}_j(t) \leq -\frac{\alpha\lambda\rho}{2}\int_0^L e^{-\lambda t}[\dot{w}_j(x,t)]^2dx - \frac{\alpha\lambda EI}{2}\int_0^L e^{-\lambda t}[w_j''(x,t)]^2dx \\ -\frac{\alpha\lambda T}{2}\int_0^L e^{-\lambda t}[w_j'(x,t)]^2dx - \alpha\Big[\frac{\lambda M}{2} - \gamma - \nu_1\bar{f}L\nu_5 - \nu_4(\nu_3u^*(1-\tanh(1)) \\ +\nu_2\bar{d})\Big]e^{-\lambda t}[\dot{w}_j(L,t)]^2 + \alpha\int_0^L e^{-\lambda t}\dot{w}_j(x,t)f_j(x,t)dx + \alpha e^{-\lambda t}\dot{w}_j(L,t)d_j(t) \\ +\alpha e^{-\lambda t}\dot{w}_j(L,t)u_{1j}(t) - \alpha\nu_1\bar{f}Le^{-\lambda t}\dot{w}_j(L_e,t)\tanh(\nu_5\dot{w}_j(L,t)) \\ -\alpha\Big[\nu_2\bar{d} + \nu_3u^*[1-\tanh(1)]\Big]e^{-\lambda t}\dot{w}_j(L,t)\tanh(\nu_4\dot{w}_j(L,t)) \\ +\alpha u^*[1-\tanh(1)]e^{-\lambda t}|\dot{w}_j(L,t)| \tag{9.22}$$

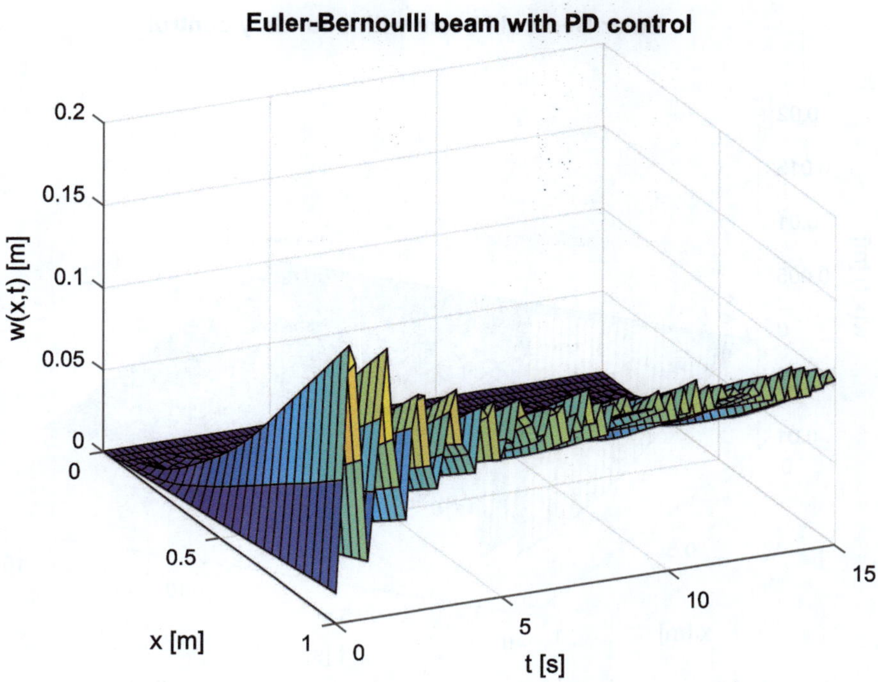

Fig. 7.3 Displacement of the flexible beam system with the PD control

To demonstrate the effectiveness and practicability of the proposed control for a classical flexible beam system with input backlash, we choose the finite difference method to simulate the system performance with boundary control. By subdividing the domain of the problem by introducing a mesh of discrete points for each of the independent variables, the finite difference method provides a straightforward and accurate process for solving PDEs involving two independent variables [9]. The central difference and explicit method are employed for simulations in this chapter.

The system given by governing equation (7.8) and boundary conditions (7.9)–(7.12) is excited by the external vary-time disturbance, and the corresponding initial conditions can be generated by the following equations

$$w(x,0) = x \tag{7.52}$$

$$\dot{w}(x,0) = 0 \tag{7.53}$$

In order to analyze the performance of the proposed control conveniently, we have considered of three different cases.

Without control: The deflection of the flexible beam system without control namely, $v(t) = 0$ under external vary-time disturbance is shown Fig. 7.2. In this case, the large vibration caused by the external disturbances would damage to the system

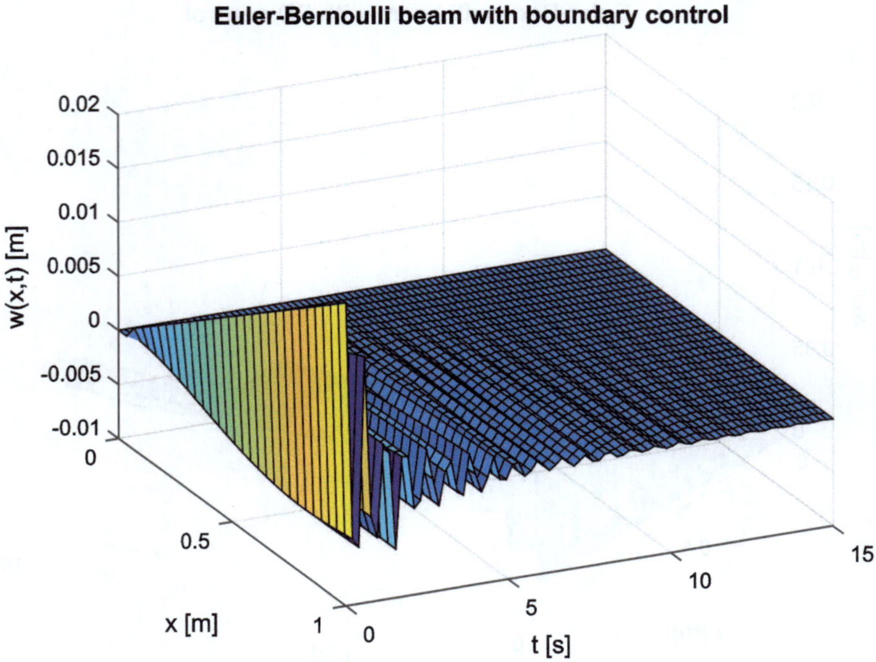

Fig. 7.4 Displacement of the flexible beam system with the proposed control

performance, such as, the undesirable inaccuracy or instability, even make the beam break.

With the PD control: The flexible beam system is analyzed with the traditional PD control $u(t) = -k_p w(L, t) - k_d \dot{w}(L, t)$ by choosing $k_p = 600$ and $k_d = 800$. Please note that, the values of k_p and k_d affect the control performance. We have tested different values of k_p and k_d ($k_p, k_d = 1, 10, 20, 50, 100, 200, 400, 600, 800, 1000$), and the best regulating performance is achieved with $k_p = 600, k_d = 800$. The transverse displacement of the beam under the PD control is shown in Fig. 7.3. We can observe that the transverse vibration has been suppressed within 5 s by the PD control.

With proposed control: The objective of boundary control is to suppress the vibration of the beam system by choosing a set of appropriate gains. In flexible beam system, we have $B_r = 90$ and $B_l = -90$. For the designed parameters, firstly, we choose $k = 100 > \frac{\beta EI}{2}$, and k_1, k_2, α and β are designed to satisfy the conditions. Then, we design $k_1 = 1, k_2 = 4$, and substitute the system parameters into the conditions, we have

Differentiating the second term in (9.10) with respect to time, we can obtain

$$\frac{d}{dt}\Big\{\frac{\alpha}{2\gamma}\int_0^t e^{-\lambda\tau}[U_j(\tau)]^2 d\tau\Big\}$$
$$\leq \frac{\alpha}{2\gamma}e^{-\lambda t}[u^*\tanh(\frac{U_{(j-1)}(t)}{u^*})]^2 - \frac{\alpha\gamma}{2}e^{-\lambda t}[\dot{w}_j(L,t)]^2 - \alpha e^{-\lambda t}\dot{w}_j(L,t)U_j(t) \tag{9.23}$$

Utilizing Young's inequality, we have

$$\begin{aligned}\dot{E}_j(t) \leq &-(\frac{\alpha\lambda\rho}{2} - \alpha\delta_1)\int_0^L e^{-\lambda t}[\dot{w}_j(x,t)]^2dx - \frac{\alpha\lambda EI}{2}\int_0^L e^{-\lambda t}[w_j''(x,t)]^2dx \\ &-\frac{\alpha\lambda T}{2}\int_0^L e^{-\lambda t}[w_j'(x,t)]^2dx + \frac{\alpha}{\delta_1}\int_0^L [f_j(x,t)]^2dx - \alpha\Big[\frac{\lambda M}{2} - \frac{\gamma}{2} \\ &-\delta_2 - \delta_3 - \nu_1\bar{f}L\nu_5 - \nu_4(\nu_2\bar{d} + \nu_3u^*(1-\tanh(1)))\Big]e^{-\lambda t}[\dot{w}_j(L,t)]^2 \\ &+\frac{\alpha}{\delta_2}[d_j(t)]^2 + \frac{\alpha}{\delta_3}[u^*[1-\tanh(1)]]^2 + \frac{\alpha}{2\gamma}e^{-\lambda t}[u^*\tanh(\frac{U_{(j-1)}}{u^*})]^2 \\ \leq &-\chi_1V_j(t) + \epsilon_j(t)\end{aligned} \tag{9.24}$$

where χ_1 and $\epsilon_j(t)$ are expressed by

$$\begin{aligned}\chi_1 = \min\{&\frac{\rho\lambda - 2\delta_1}{\rho}, [\frac{M\lambda - \gamma - 2\delta_2 - 2\delta_3}{M} - \frac{2\nu_1\bar{f}L\nu_4 + 2\nu_4\nu_2\bar{d}}{M} \\ &-\frac{2\nu_4\nu_3u^*(1-\tanh(1))}{M}]\} > 0\end{aligned} \tag{9.25}$$

$$\begin{aligned}\epsilon_j(t) = &\frac{\alpha}{2\gamma}e^{-\lambda t}[u^*\tanh(\frac{U_{(j-1)}(t)}{u^*})]^2 + \frac{\alpha}{\delta_1}\int_0^L [f_j(x,t)]^2dx + \frac{\alpha}{\delta_2}[d_j(t)]^2 \\ &+\frac{\alpha}{\delta_3}[(1-\tanh(1))u^*]^2 \leq \bar{\epsilon}\end{aligned} \tag{9.26}$$

where δ_1–δ_3 are three positive constants. $\bar{\epsilon}$ is a positive constant, satisfying $\bar{\epsilon} = \frac{\alpha}{2\gamma}[u^*]^2 + \frac{\alpha\bar{f}^2L}{\delta_1} + \frac{\alpha\bar{d}^2}{\delta_2} + \frac{\alpha[(1-\tanh(1))u^*]^2}{\delta_3}$. Therefore, we can obtain $V_j(t) \leq V_j(0)e^{-\chi_1 t} + \frac{\bar{\epsilon}}{\chi_1}$, which advises the boundedness of the closed-loop Euler-Bernoulli beam system.

Step II: Boundedness of $E_0(t)$

Proof Considering Assumption 9.1, the difference of $E_j(T_b)$ is expressed by

$$\Delta E_j(T_b) \leq -\frac{\alpha\lambda\rho}{2}\int_0^{T_b}\int_0^L e^{-\lambda\tau}[\dot{w}_{ej}(x,r)]^2 dxd - \frac{\alpha\lambda EI}{2}\int_0^{T_b}\int_0^{L_e} e^{-\lambda\tau}[w_j''(x,\tau)]^2 dxd\tau$$
$$-\frac{\alpha\lambda T}{2}\int_0^{T_b}\int_0^L e^{-\lambda\tau}[w_j'(x,\tau)]^2 dxd\tau + \eta(1-e^{-\lambda T_b})\int_0^{T_b} V_j(\tau)d\tau$$
$$-\alpha\Big[\frac{\lambda M}{2} - \frac{\gamma}{2} - \nu_1\bar{f}L\nu_5 - \nu_4(\nu_2\bar{d} + \nu_3 u^*(1-\tanh(1)))\Big]\int_0^{T_b} e^{-\lambda\tau}[\dot{w}_j(L,\tau)]^2 d\tau$$
$$+\alpha\int_0^{T_b}\int_0^L e^{-\lambda\tau}\dot{w}_j(x,\tau)f_j(x,\tau)dxd\tau + \alpha u^*[1-\tanh(1)]\int_0^{T_b} e^{-\lambda\tau}|\dot{w}_j(L,\tau)|d\tau$$
$$+\alpha\int_0^{T_b} e^{-\lambda\tau}\dot{w}_j(L,\tau)d_j(r)d\tau - \alpha\int_0^{T_b}\nu_1\bar{f}Le^{-\lambda\tau}\dot{w}_j(L,\tau)\tanh(\nu_5\dot{w}_j(L,\tau))d\tau$$
$$-\alpha\Big[\nu_2\bar{d} + \nu_3 u^*[1-\tanh(1)]\Big]\int_0^{T_b} e^{-\lambda\tau}\dot{w}_j(L,\tau)\tanh(\nu_4\dot{w}_j(L,\tau))d\tau \quad (9.27)$$

$E_n(t)$ is then expressed by

$$E_n(T_b) \leq -(\chi_2 - \eta_e(1-e^{-\lambda T_b}))\sum_{i=1}^{i=n}\int_0^{T_b} V_i(\tau)d\tau + G_{u,n}(T_b) + G_{d,n}(T_b) + G_{f,n}(T_b)$$
$$+E_{e0}(T_b) \quad (9.28)$$

where $\chi_2 - \eta(1-e^{-\lambda T_b}) > 0$ and $\chi_2 = \frac{1}{M}[\lambda M - \gamma - 2\nu_1\bar{f}L\nu_5 - 2\nu_4(\nu_2\bar{d} + \nu_3 u^*(1-\tanh(1)))] > 0$.

The auxiliary functions $G_{u,n}(T_b)$, $G_{d,n}(T_b)$ and $G_{f,n}(T_b)$ are defined as

$$G_{u,n}(t) = \alpha u^*[1-\tanh(1)]\sum_{k=1}^{k=N_1}\sum_{i=q_k}^{i=q_{k+1}}\int_0^{T_b} e^{-\lambda\tau}\Big[|\dot{w}_i(L,\tau)| - \nu_3\dot{w}_i(L,\tau)$$
$$\times\tanh(\nu_4\dot{w}_i(L,\tau))\Big]d\tau \quad (9.29)$$

$$G_{d,n}(t) = \alpha\sum_{k=1}^{k=N_1}\sum_{i=q_k}^{i=q_{k+1}}\int_0^{T_b} e^{-\lambda t}\dot{w}_i(L,\tau)\Big[d_i(\tau) - \nu_2\bar{d}\tanh(\nu_4\dot{w}_i(L,\tau))\Big]d\tau \quad (9.30)$$

$$G_{f,n}(t) = \alpha\sum_{k=1}^{k=N_2}\sum_{i=p_k}^{i=p_{k+1}}\int_0^{T_b} e^{-\lambda\tau}\Big\{\int_0^L \dot{w}_j(x,\tau)f_j(x,\tau)dx$$
$$-\nu_1\bar{f}Le^{-\lambda\tau}\dot{w}_i(L,\tau)\tanh(\nu_5\dot{w}_i(L,\tau))\Big\}d\tau \quad (9.31)$$

where the interval sequences are defined below.

In order to easily find ν_2 and ν_4, $G_d(t)$ is rewritten as

$$G_d(t) \leq \alpha\bar{d}\sum_{k=1}^{k=N_1}\sum_{i=q_k}^{q_{k+1}}\int_0^{T_b} e^{-\lambda r}\Big\{|\dot{w}_i(L,\tau)| - \nu_2\dot{w}_i(L,\tau)\tanh(\nu_4\dot{w}_i(L,\tau))\Big\}d\tau$$
$$(9.32)$$

If finite ν_2 and ν_4 can be found to make the right side of the above inequality less than zero, they also satisfy $G_d(t) \leq 0$.

Likewise,

$$G_f(t) \leq \alpha \bar{f} \sqrt{L} \sum_{k=1}^{k=N_2} \sum_{i=p_k}^{i=p_{k+1}} \int_0^{T_b} e^{-\lambda\tau} \left\{ \sqrt{\int_0^L [\dot{w}_i(x,\tau)]^2 dx} \right.$$

$$\left. -\nu_1 \sqrt{L} \dot{w}_i(L,\tau) \tanh(\nu_5 \dot{w}_i(L,\tau)) \right\} d\tau \qquad (9.33)$$

The above interval sequences should satisfy

(I): $\bigcup_{k=\{1,2,\cdots,N_1\}} [(q_k - 1)T_b, q_{k+1}T_b] = \bigcup_{i=\{1,2,\cdots,n\}} [0, T_b]$ and similarly $\bigcup_{k=\{1,2,\cdots,N_2\}} [(p_k - 1)T_b, p_{k+1}T_b] = \bigcup_{i=\{1,2,\cdots,n\}} [0, T_b]$, where $0 < q_k < q_{k+1} < +\infty$ and $0 < q_k < q_{k+1} < +\infty$ for $k \in \mathbb{N}$.

(II): N_1 and N_2 would go to infinity for $n \to \infty$.

(III): $[(q_k - 1)T_b, q_{k+1}T_b] \supsetneqq [t_{q,k}, t_{q,k+1}]$, where $\sup_{t \in [t_{q,k}, t_{q,k+1}]} |\dot{w}_j(L,t)| < \max\{\frac{1}{\nu_4} \tanh^{-1}(\frac{1}{\nu_2}), \frac{1}{\nu_4} \tanh^{-1}(\frac{1}{\nu_3})\}$, and otherwise not.

(IV): Likewise, $[(p_k - 1)T_b, p_{k+1}T_b] \supsetneqq [t_{p,k}, t_{p,k+1}]$, where for $\forall t \in [t_{p,k}, t_{p,k+1}]$, $\nu_1 \sqrt{L} \dot{w}_j(L,t) \tanh(\nu_5 \dot{w}_j(L,t)) < \sqrt{\int_0^L [\dot{w}_j(x,t)]^2 dx}$, and otherwise not.

Such interval sequences satisfying (I)–(IV), are defined reasonably. It is easy to understand (I) and (II). If $G_d(t) < 0$, $G_u(t) < 0$ and $G_f(t) < 0$, the external disturbances and the input nonlinearity cannot completely addressed. This would consequently trigger the vibrations and finally make $[t_{q,k}, t_{q,k+1}]$ and $[t_{q,k}, t_{q,k+1}]$ finite. Then, finite $[(q_k - 1)T_b, q_{k+1}T_b]$ and $[(p_k - 1)T_b, p_{k+1}T_b]$ can be found to contain them, respectively.

By choosing proper ν_1–ν_5, $E_n(T_b)$ is then updated as

$$\lim_{n\to\infty} E_n(T_b) \leq E_0(T_b) - (\chi_2 - \eta_e(1 - e^{-\lambda T_b})) \sum_{i=1}^{+\infty} \int_0^{T_b} V_i(\tau) d\tau \qquad (9.34)$$

In order to satisfy the positiveness of $E_n(T_b)$, $\forall j \in \mathbb{N}$, the series $\sum_{i=1}^{i=n} \int_0^{T_b} V_i(\tau) d\tau$ should be not be infinite with $n \to +\infty$. In other words, $\lim_{j\to\infty} \int_0^{T_b} V_j(\tau) d\tau = 0$. Considering (9.11) and utilizing Lemma in [44], we further prove that $w_j(x,t)$, $w'_j(x,t)$, $w''_j(x,t)$ and $\dot{w}_j(x,t)$ converge to zero for $\forall (x,t) \in [0, L] \times [0, T_b]$, as $j \to +\infty$.

9.5 Numerical Simulations

In this section, a simulation example is presented to illustrate the effectiveness of the proposed ABILC scheme (9.6). The distributed disturbances and the boundary disturbances are assumed as

Table 9.1 Parameters of the Euler-Bernoulli beam system

Parameter	Description	Value
L	Length of the beam	1 m
ρ	Unit mass per unit length of the beam	0.1 kg/m
EI	Bending stiffness of the beam	$6\,\text{Nm}^2$
T	Tension of the beam	10 N
M	Mass of the tip payload	1 kg

$$f_j(x,t) = \frac{x}{20}[\sin(0.1j\pi xt) + \sin(0.2j\pi xt) + \sin(0.3j\pi xt)], \tag{9.35}$$

$$d_j(t) = \frac{1}{10}[\sin(j\pi t) + \sin(2j\pi t) + \sin(3j\pi t)], \tag{9.36}$$

where $T_b = 5\,\text{s}$. The corresponding initial conditions are $w_j(x,0) = 0.8x\,\text{m}$ and $\dot{w}_j(x,0) = 0.1\,\text{m/s}$. The iteration number is $j = \{0, 1, 2, \ldots, 28, 29\}$. We also set $u^* = 15\,\text{N}$. The parameters of the Euler-Bernoulli beam system are given in Table 9.1.

Without control, under the aperiodic distributed disturbance and the aperiodic boundary disturbance, the displacements repeatedly go up to 1 m and down to −1 m in Fig. 9.1. From Fig. 9.2, the endpoint vibrations vary largely and quickly. For the Euler-Bernoulli beam of $L = 1$ m, such large transverse displacements would degrade the Euler-Bernoulli beam system performance and even give rise to the system paralysis.

By employing the ABILC law with $\nu_1 = 1, \nu_2 = \nu_3 = 2, \nu_4 = 5$, and $\gamma = 200$, the closed-loop system is described through Figs. 9.3, 9.4, 9.5, 9.6 and 9.7. As shown in Fig. 9.3, the transverse displacements of the Euler-Bernoulli beam are asymptotically suppressed to zero within 100 s. From Fig. 9.4, the vibrations in the final iteration cycle are suppressed into an acceptable region around zero, namely, $|w_{29}(x,t)| \leq 0.002$ m through the proposed ABILC scheme. By comparing Fig. 9.2 with Fig. 9.5, $w_j(L,t)$ of the closed-loop Euler-Bernoulli beam system converges to zero as time increases. Along the iteration axis, the maximum absolute values of the displacements reduce dramatically within 20 iterations in Fig. 9.6. The control input in Fig. 9.7 is restrained with $[-50\,\text{N}, 50\,\text{N}]$.

Figure 9.3 and Fig. 9.5 present the convergence of $w_j(x,t)$ and $w_j(L,t)$ along the time axis. Figure 9.4 is used to illustrate the boundedness of $w_j(x,t)$ in the final iteration. Figure 9.6 reveals the effectiveness of the proposed ABILC law along the iteration axis. Therefore, from the above simulation results, the displacements

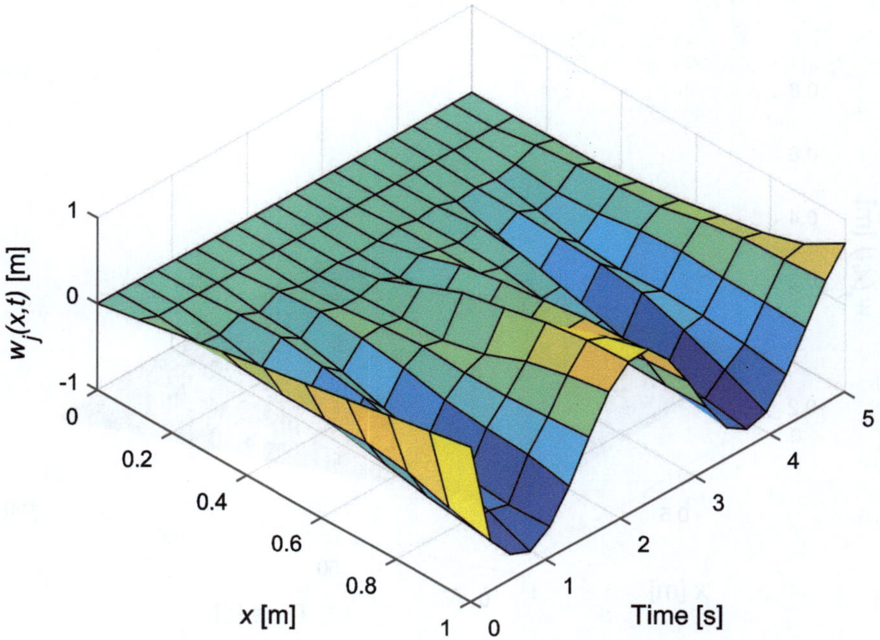

Fig. 9.1 $w_j(x, t)$ without control

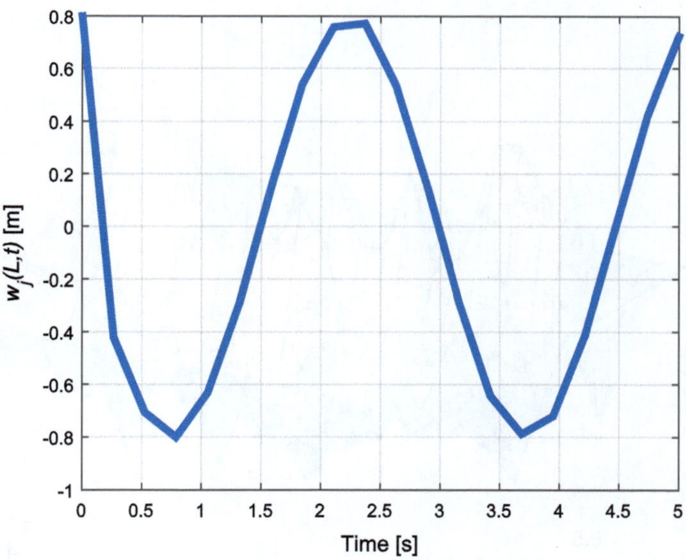

Fig. 9.2 $w_j(L, t)$ without control

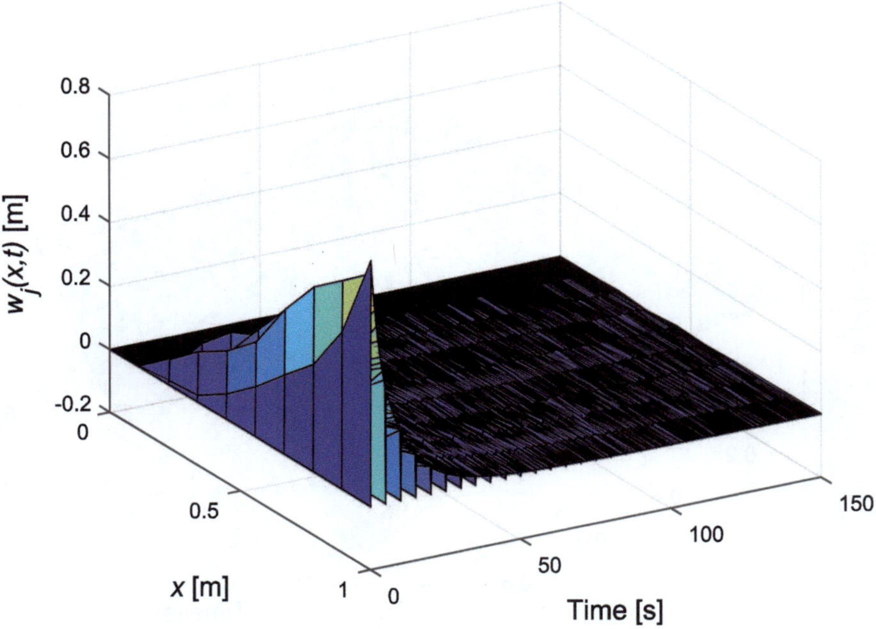

Fig. 9.3 $w_j(x, t)$ with the proposed ABILC law

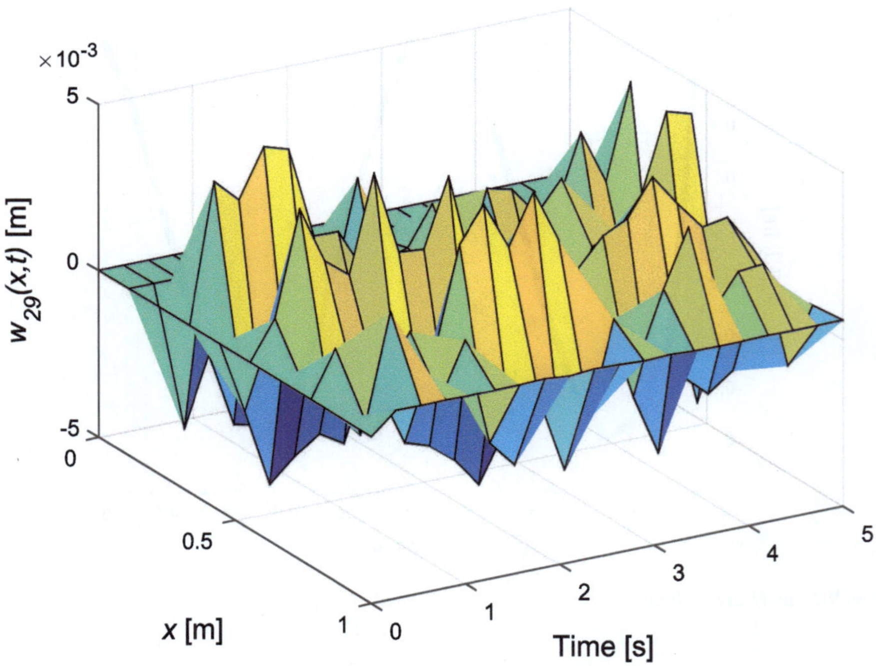

Fig. 9.4 $w_{29}(x, t)$ with the proposed ABILC law

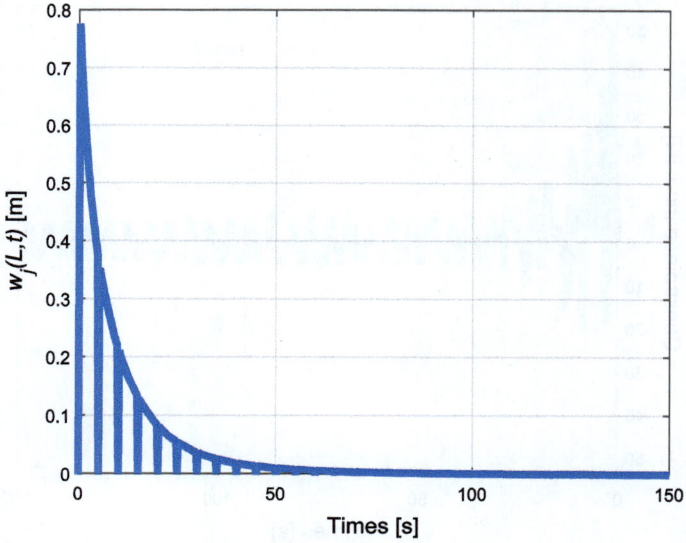

Fig. 9.5 $w_j(L, t)$ with the proposed ABILC law

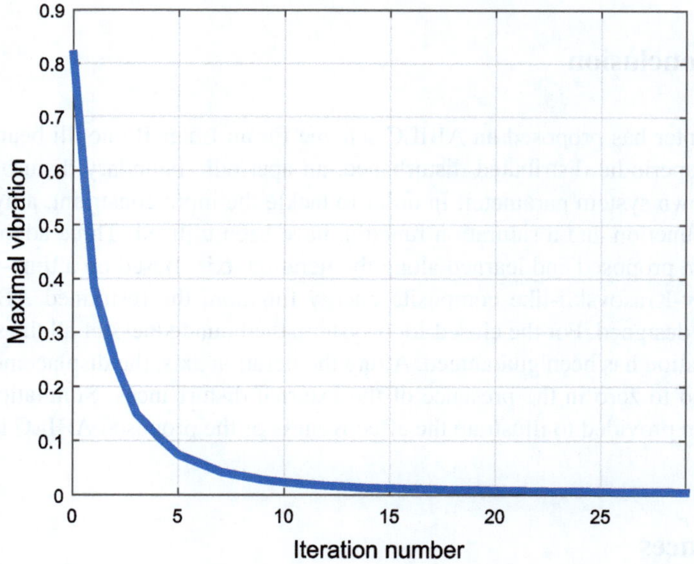

Fig. 9.6 $\max\{|w_j(x, t)|\}$ with the proposed ABILC law

converge to zero along the iteration axis and the input constraint is tackled for the Euler-Bernoulli beam in the presence of the external disturbances and the uncertainty of the system parameter.

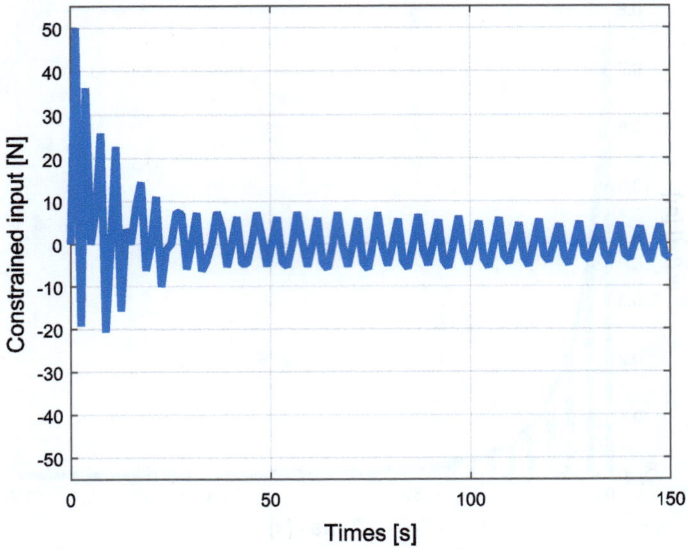

Fig. 9.7 Control input $\bar{u}\text{sat}(\frac{u_j(t)}{\bar{u}}, 1)$

9.6 Conclusion

This chapter has proposed an ABILC scheme for an Euler-Bernoulli beam system with an aperiodic distributed disturbance, an aperiodic boundary disturbance and an unknown system parameter. In order to tackle the input constraint, a hyperbolic tangent function and a saturation function have been utilized. Three adaptive laws have been proposed and learned along the iteration axis. Based on a time-weighted Lyapunov-Krasovskii-like composite energy function, the restrained ABILC law has been designed. For the closed-loop system, the boundedness of all the signals in each iteration has been guaranteed. Along the iteration axis, the displacements have converged to zero in the presence of the external disturbances. Simulation results have been provided to illustrate the effectiveness of the proposed ABILC law.

References

1. X. Li, J.-X. Xu, D. Huang, An iterative learning control approach for linear time-invariant systems with randomly varying trial lengths. IEEE Trans. Autom. Control **59**(7), 1954–1960 (2014)
2. R. Chi, Z. Hou, S. Jin, D. Wang, C.-J. Chien, Enhanced data-driven optimal terminal ILC using current iteration control knowledge. IEEE Trans. Neural Netw. Learn. Syst. **26**(11), 2939–2948 (2015)
3. D. Liu, D. Wang, H. Li, Decentralized stabilization for a class of continuous-time nonlinear interconnected systems using online learning optimal control approach. IEEE Trans. Neural

Netw. Learn. Syst. **25**(2), 418–428 (2014)
4. X. Li, D. Huang, B. Chu, J.-X. Xu, Robust iterative learning control for systems with norm-bounded uncertainties. Int. J. Robust Nonlinear Control **26**(4), 697–718 (2016)
5. M.-B. Radac, R.-E. Precup, E.M. Petriu, Model-free primitive-based iterative learning control approach to trajectory tracking of MIMO systems with experimental validation. IEEE Trans. Neural Netw. Learn. Syst. **26**(11), 2925–2938 (2015)
6. D. Meng, Y. Jia, J. Du, Robust consensus tracking control for multiagent systems with initial state shifts, disturbances, and switching topologies. IEEE Trans. Neural Netw. Learn. Syst. **26**(4), 809–824 (2015)
7. P. Zanchetta, M. Degano, J. Liu, P. Mattavelli, Iterative learning control with variable sampling frequency for current control of grid-connected converters in aircraft power systems. IEEE Trans. Ind. Appl. **49**(4), 1548–1555 (2013)
8. D. Huang, J.-X. Xu, X. Li, C. Xu, M. Yu, D-type anticipatory iterative learning control for a class of inhomogeneous heat equations. Automatica **49**(8), 2397–2408 (2013)
9. F.-M. Chen, J.S.-H. Tsai, Y.-T. Liao, S.-M. Guo, M.-C. Ho, F.-Z. Shaw, L.-S. Shieh, An improvement on the transient response of tracking for the sampled-data system based on an improved PD-type iterative learning control. J. Frankl. Inst. **351**(2), 1130–1150 (2014)
10. F. Ma, C. Li, Open-closed-loop PID-type iterative learning control for linear systems with initial state error. J. Vib. Control **17**(12), 1791–1797 (2011)
11. R. Chi, Z. Hou, J. Xu, Adaptive ILC for a class of discrete-time systems with iteration-varying trajectory and random initial condition. Automatica **44**(8), 2207–2213 (2008)
12. J.-X. Xu, B. Viswanathan, Adaptive robust iterative learning control with dead zone scheme. Automatica **36**(1), 91–99 (2000)
13. A. Tayebi, Adaptive iterative learning control for robot manipulators. Automatica **40**(7), 1195–1203 (2004)
14. M. Norrlöf, An adaptive iterative learning control algorithm with experiments on an industrial robot. IEEE Trans. Robot. Autom. **18**(2), 245–251 (2002)
15. A. Tayebi, C.-J. Chien, A unified adaptive iterative learning control framework for uncertain nonlinear systems. IEEE Trans. Autom. Control **52**(10), 1907–1913 (2007)
16. D. Huang, X. Li, J.-X. Xu, C. Xu, W. He, Iterative learning control of inhomogeneous distributed parameter systemsfrequency domain design and analysis. Syst. Control Lett. **72**, 22–29 (2014)
17. Z. Qu, An iterative learning algorithm for boundary control of a stretched moving string. Automatica **38**(5), 821–827 (2002)
18. H. Zhao, C.D. Rahn, Iterative learning velocity and tension control for single span axially moving materials. J. Dyn. Syst. Meas. Control **130**(5), 051003 (2008)
19. M. Krstic, A. Smyshlyaev, Adaptive boundary control for unstable parabolic PDEs-Part I: Lyapunov design. IEEE Trans. Autom. Control **53**(7), 1575–1591 (2008)
20. J.-W. Wang, H.-N. Wu, H.-X. Li, Stochastically exponential stability and stabilization of uncertain linear hyperbolic PDE systems with markov jumping parameters. Automatica **48**(3), 569–576 (2012)
21. B. Luo, H.-N. Wu, H.-X. Li, Adaptive optimal control of highly dissipative nonlinear spatially distributed processes with neuro-dynamic programming. IEEE Trans. Neural Netw. Learn. Syst. **26**(4), 684–696 (2015)
22. S. Zhang, W. He, D. Huang, Active vibration control for a flexible string system with input backlash. IET Control Theor. Appl. **10**(7), 800–805 (2016)
23. W. He, S. Zhang, S.S. Ge, Adaptive boundary control of a nonlinear flexible string system. IEEE Trans. Control Syst. Technol. **22**(3), 1088–1093 (2014)
24. B.-Z. Guo, F.-F. Jin, Output feedback stabilization for one-dimensional wave equation subject to boundary disturbance. IEEE Trans. Autom. Control **60**(3), 824–830 (2015)
25. B.-Z. Guo, H.-C. Zhou, The active disturbance rejection control to stabilization for multi-dimensional wave equation with boundary control matched disturbance. IEEE Trans. Autom. Control **60**(1), 143–157 (2015)
26. W. He, S. Nie, T. Meng, Y.-J. Liu, Modeling and vibration control for a moving beam with application in a drilling riser. IEEE Trans. Control Syst. Technol. **25**(3), 1036–1043 (2017)

27. H.-N. Wu, J.-W. Wang, Static output feedback control via PDE boundary and ODE measurements in linear cascaded ODE-beam systems. Automatica **50**(11), 2787–2798 (2014)
28. D. Liu, L. Zhang, G. Xu, Stabilisation of Timoshenko beam system with a tip payload under the unknown boundary external disturbances. Int. J. Control **88**(9), 1830–1840 (2015)
29. S. Zhang, D. Huang, End-point regulation and vibration suppression of a flexible robotic manipulator. Asian J. Control **19**(1), 245–254 (2017)
30. W. He, S.S. Ge, Robust adaptive boundary control of a vibrating string under unknown time-varying disturbance. IEEE Trans. Control Syst. Technol. **20**(1), 48–58 (2012)
31. W. He, X. He, S.S. Ge, Vibration control of flexible marine riser systems with input saturation. IEEE/ASME Trans. Mechatron. **21**(1), 254–265 (2016)
32. W. He, S.S. Ge, D. Huang, Modeling and vibration control for a nonlinear moving string with output constraint. IEEE/ASME Trans. Mechatron. **20**(4), 1886–1897 (2015)
33. J.-K. Park, C.-H. Choi, H. Choo, Dynamic anti-windup method for a class of time-delay control systems with input saturation. Int. J. Robust Nonlinear Control **10**(6), 457–488 (2000)
34. Y. Li, S. Tong, T. Li, Adaptive fuzzy output-feedback control for output constrained nonlinear systems in the presence of input saturation. Fuzzy Sets Syst. **248**, 138–155 (2014)
35. G. Grimm, J. Hatfield, I. Postlethwaite, A.R. Teel, M.C. Turner, L. Zaccarian, Antiwindup for stable linear systems with input saturation: an lmi-based synthesis. IEEE Trans. Autom. Control **48**(9), 1509–1525 (2003)
36. C. Wen, J. Zhou, Z. Liu, H. Su, Robust adaptive control of uncertain nonlinear systems in the presence of input saturation and external disturbance. IEEE Trans. Autom. Control **56**(7), 1672–1678 (2011)
37. S.S. Ge, S. Zhang, W. He, Vibration control of an Euler-Bernoulli beam under unknown spatiotemporally varying disturbance. Int. J. Control **84**(5), 947–960 (2011)
38. J.-X. Xu, Y. Tan, T.-H. Lee, Iterative learning control design based on composite energy function with input saturation. Automatica **40**(8), 1371–1377 (2004)
39. L. Sun, W. Huo, Z. Jiao, Adaptive backstepping control of spacecraft rendezvous and proximity operations with input saturation and full-state constraint. IEEE Trans. Ind. Electron. **64**(1), 480–492 (2017)
40. Q. Zhou, L. Wang, C. Wu, H. Li, H. Du, Adaptive fuzzy control for non-strict feedback systems with input saturation and output constraint. IEEE Trans. Syst. Man Cybern. Syst. **47**(1), 1–12 (2017)
41. D. Huang, J.-X. Xu, Steady-state iterative learning control for a class of nonlinear PDE processes. J. Process Control **21**(8), 1155–1163 (2011)
42. W. He, T. Meng, D. Huang, X. Li, Adaptive boundary iterative learning control for an Euler-Bernoulli beam system with input constraint. IEEE Trans. Neural Netw. Learn. Syst. **29**(5), 1539–1549 (2018)
43. T. Meng, W. He, Iterative learning control of a robotic arm experiment platform with input constraint. IEEE Trans. Ind. Electron. **65**(1), 664–672 (2017)
44. G.H. Hardy, J.E. Littlewood, G. Polya, *Inequalities* (Cambridge University Press, Cambridge, UK, 1959)

Chapter 10
Neural Network Control of a Flexible Beam

10.1 Introduction

In recent years, control theory and method has been widely used in many fields and has been made great progress in the research of adaptive control [1–3]. Among the research methods, the adaptive control [4, 5] technology provides a powerful tool to solve the model uncertainty caused by the variety of parameters. The design of the adaptive controller is based on the identification of the system parameters. However, the identification and computation of complex system are time-consuming, which makes it difficult to realize real-time control of the fast system. In recent years, artificial intelligence [6], especially the research of neural network (NN) [7–9], has provided an effective method to solve these problems.

Neural networks based approach is considered to be an effective method in a number of research fields, which requires relatively less information of the system dynamics. It has been proven that artificial neural networks are able to approximate a wide range of nonlinear functions to any desired degree of accuracy under certain conditions [10, 11]. Artificial neural networks have been widely used for the control design of uncertain nonlinear systems [12–15]. There are many relevant applications for this approach based on the Lyapunov stability theory [16–20].

For neural networks controller design in general, the first step is to establish an approximate dynamic model [21–23], then an appropriate control strategy that utilizes this approximation can be constructed. This approach works well for many systems.

In the assumed mode method (AMM) [24–26] modeling, the elastic deflection of the flexible beam is represented by, theoretically an infinite number of separable modes, but practically only finite number of modes with comparatively low frequencies are considered as they are generally dominant in the system's dynamic behavior. The method of arc approximation is used to represent the position of the flexible link, which leads to a linear time-invariant model.

W. He and J. Liu, *Active Vibration Control and Stability Analysis of Flexible Beam Systems*, https://doi.org/10.1007/978-981-10-7539-1_10

In this chapter, we investigate an Euler–Bernoulli beam model via AMM. Based on the N-dimensional model, the neural network control is proposed to achieve vibration suppression.

The rest of the chapter is arranged as follows. In Sect. 10.2, the model of the Euler–Bernoulli beam system is given via AMM. Neural network control scheme with both full-state feedback and output feedback is designed in Sect. 10.3. Simulations are provided to illustrate the effectiveness of the proposed NN law in Sect. 10.4. The conclusion follows in Sect. 10.5.

10.2 Problem Formulation

Figure 10.1 shows an Euler–Bernoulli beam model. The left boundary of the beam is fixed at origin.

Hamilton principle which is used to derive the model of the beam system is represented by

$$\int_{t_1}^{t_2} \delta[E_k(t) - E_p(t) + W(t)]dt = 0, \tag{10.1}$$

where t_1 and t_2 are two time instants, $t_1 < t < t_2$ is the operating interval and δ denotes the variational operator.

The kinetic energy of the beam $E_k(t)$ can be represented as

$$E_k(t) = \frac{1}{2}M[\dot{\omega}(L,t)]^2 + \frac{1}{2}\rho\int_0^L [\dot{\omega}(x,t)]^2 dx, \tag{10.2}$$

where x and t represent the independent spatial and time variables, respectively.

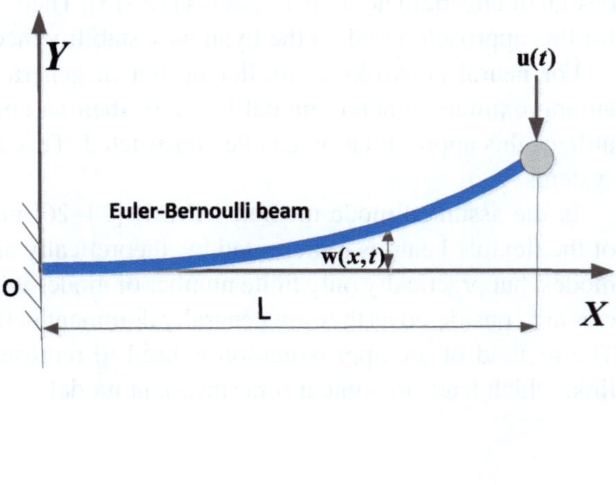

Fig. 10.1 A typical flexible cantilevered beam system

The potential energy $E_p(t)$ due to the bending can be obtained from

$$E_p(t) = \frac{1}{2} EI \int_0^L [\omega''(x,t)]^2 dx, \tag{10.3}$$

The virtual work done by the control input force $u(t)$ which produces a transverse force for vibration suppression can be written as

$$\delta W(t) = u(t)\delta\omega(L,t), \tag{10.4}$$

Applying the variation operator and integrating (10.2), (10.3), (10.4) by parts respectively, we obtain

$$\begin{aligned}
\int_{t_1}^{t_2} \delta E_k(t)dt &= M\int_{t_1}^{t_2} \dot{\omega}(L,t)\delta\dot{\omega}(L,t)dt + \rho\int_{t_1}^{t_2}\int_0^L \dot{\omega}(x,t)\delta\dot{\omega}(x,t)dxdt \\
&= M\dot{\omega}(L,t)\delta\omega(L,t) \mid_{t_1}^{t_2} - M\int_{t_1}^{t_2} \ddot{\omega}(L,t)\delta\omega(L,t)dt \\
&\quad +\rho\int_0^L [\dot{\omega}(x,t)\delta\omega(x,t) \mid_{t_1}^{t_2}]dx - \rho\int_0^L\int_{t_1}^{t_2} \ddot{\omega}(x,t)\delta\omega(x,t)dtdx \\
&= -M\int_{t_1}^{t_2} \ddot{\omega}(L,t)\delta\omega(L,t)dt - \rho\int_0^L\int_{t_1}^{t_2} \ddot{\omega}(x,t)\delta\omega(x,t)dtdx. \quad (10.5) \\
\int_{t_1}^{t_2} \delta E_p(t)dt &= \int_{t_1}^{t_2}\int_0^L [EI\omega''(x,t)\delta\omega''(x,t)]dxdt, \\
&= \int_{t_1}^{t_2} [EI\omega''(x,t)\delta\omega'(x,t) \mid_0^L - EI\omega'''(x,t)\delta\omega(x,t) \mid_0^L \\
&\quad + \int_0^L EI\omega''''(x,t)\delta\omega(x,t)dx]dt. \quad (10.6) \\
\int_{t_1}^{t_2} \delta W(t)dt &= \int_{t_1}^{t_2} u(t)\delta\omega(L,t)dt. \quad (10.7)
\end{aligned}$$

Substituting (10.5), (10.6) and (10.7) into Hamiltons principle (10.1), we obtain

$$\begin{aligned}
&-M\int_{t_1}^{t_2} \ddot{\omega}(L,t)\delta\omega(L,t)dt - \rho\int_0^L\int_{t_1}^{t_2} \ddot{\omega}(x,t)\delta\omega(x,t)dtdx \\
&-\int_{t_1}^{t_2} [EI\omega''(x,t)\delta\omega'(x,t) \mid_0^L - EI\omega'''(x,t)\delta\omega(x,t) \mid_0^L \\
&+\int_0^L EI\omega''''(x,t)\delta\omega(x,t)dx]dt + \int_{t_1}^{t_2} u(t)\delta\omega(L,t)dt = 0. \quad (10.8)
\end{aligned}$$

As $\delta\omega(x,t)$ is assumed to be an nonzero arbitrary variations in $0 < x < L$, the expressions under the double integral in (10.8) are set equal to zero. Hence, we obtain the governing equations of the system as

$$\rho\ddot{\omega}(x,t) + EI\omega''''(x,t) = 0, \tag{10.9}$$

where $\forall(x,t) \in (0,L) \times [0,\infty)$. Setting the terms with single integrals in (10.8) equal to zero, we obtain the boundary conditions of the system as

$$\omega(0,t) = \omega'(0,t) = \omega''(L,t) = 0, \tag{10.10}$$

$$M\ddot{\omega}(L,t) - EI\omega'''(L,t) = u(t). \tag{10.11}$$

where $\forall t \in [0,\infty)$.

The deflection of the flexible link, $w(x,t)$, is described by a number of nodes, which are separable and harmonic

$$\omega_i(x,t) = \sum_{i=1}^{\infty} \phi_i(x)q_i(t). \tag{10.12}$$

where $\phi_i(x)$ denotes the ith space dependent mode shape function and $q_i(t)$ represents the ith time dependent harmonic mode.

It is well known that the first several modes (corresponding to lower frequencies) are dominant in describing the system dynamics. The infinite series can be truncated into a finite one, i.e.,

$$\omega_i(x,t) = \sum_{i=1}^{N} \phi_i(x)q_i(t), \; 0 \le x \le L. \tag{10.13}$$

where N is the number of the modes which are taken into consideration.

The assumed mode method is a discrete method for continuous systems, which uses the linear sum of assumed modes to describe the vibration of the elastic body.

Because w_1 is the dominant one in the flexible manipulator, we use the first mode to approximate the elastic deflection. where $q(t) = q_1(t)$ is the dominant elastic mode.

$$\omega(x,t) = \phi(x)q(t). \tag{10.14}$$

Substituting this solution into (10.9) yields

$$\frac{\phi''''}{\phi}\frac{EI}{\rho} = -\frac{\ddot{q}}{q}. \tag{10.15}$$

Therefore, we define a constant k. We can obtain

$$\ddot{q}(t) + kq(t) = 0, \tag{10.16}$$
$$\phi''''(x) = \frac{\rho}{EI} k\phi(x). \tag{10.17}$$

and the boundary conditions are reduced to

$$\phi(0) = \phi'(0) = \phi''(L) = 0, \tag{10.18}$$
$$M\phi(L)\ddot{q}(t) - EI\phi'''(L)q(t) = u(t). \tag{10.19}$$

Clearly, $q(t)$ and $\phi(x)$ of (10.16) and (10.17) should be such that $\omega(x,t) = \phi(x)q(t)$ satisfies the boundary conditions in (10.10)–(10.11).

Let k be a positive number given by $k = h^2$, Eq. (10.17) can be rewritten as

$$\phi''''(x) = \left(\frac{\beta}{L}\right)^4 \phi(x). \tag{10.20}$$

Consider the general solution

$$\phi(x) = C_1 \cos\frac{\beta x}{L} + C_2 \cosh\frac{\beta x}{L} + C_3 \sin\frac{\beta x}{L} + C_4 \sinh\frac{\beta x}{L}. \tag{10.21}$$

From the boundary conditions, we have the set of equations

$$C_1 + C_2 = 0, \tag{10.22}$$
$$C_3 + C_4 = 0, \tag{10.23}$$
$$-C_1 \cos\beta + C_2 \cosh\beta - C_3 \sin\beta + C_4 \sinh\beta = 0, \tag{10.24}$$
$$C_1\left(\sin\beta + \frac{M}{\rho L}\cos\beta\right) + C_2\left(\sinh\beta + \frac{M}{\rho L}\cosh\beta\right)$$
$$+C_3\left(\frac{M}{\rho L}\sin\beta - \cos\beta\right) + C_4\left(\cosh\beta + \frac{M}{\rho L}\sinh\beta\right) = 0 \tag{10.25}$$

where β denotes the minimum positive solution of the following equation

$$1 + \cosh\beta\cos\beta + \frac{M\beta}{\rho L}(\sinh\beta\cos\beta - \cosh\beta\sin\beta) = 0 \tag{10.26}$$

and the boundary conditions are rewritten as

$$\phi''''(x) = \left(\frac{\beta}{L}\right)^4 \phi(x), \tag{10.27}$$
$$\phi(0) = \phi'(0) = \phi''(L) = 0, \tag{10.28}$$

$$\phi'''(L) + \frac{M}{EI} h^2 \phi(L) = 0. \tag{10.29}$$

where β/L is given by

$$\left(\frac{\beta}{L}\right)^4 = \frac{\rho}{EI} h^2. \tag{10.30}$$

Consider the first two equations in (10.22)–(10.25), we have

$$\phi(x) = C_2 \left[\cosh \frac{\beta x}{L} - \cos \frac{\beta x}{L}\right] + C_4 \left[\sinh \frac{\beta x}{L} - \sin \frac{\beta x}{L}\right]. \tag{10.31}$$

and from the third equation in (10.22)–(10.25), $\phi(x)$ can be further written as

$$\phi(x) = C_2 \left[\cosh \frac{\beta x}{L} - \cos \frac{\beta x}{L} - \gamma \left(\sinh \frac{\beta x}{L} - \sin \frac{\beta x}{L}\right)\right]. \tag{10.32}$$

where

$$\gamma = \frac{\cosh \beta + \cos \beta}{\sinh \beta + \sin \beta} \tag{10.33}$$

Let $0 < \beta_1 < \beta_2 < \cdots < \infty$ be the infinite number of positive solutions, we can obtain an infinite number of solutions of the boundary value problem

$$\phi_i(x) = A_i \bar{\phi}_i(x) = A_i \left[\cosh \frac{\beta_i x}{L} - \cos \frac{\beta_i x}{L} - \gamma_i \left(\sinh \frac{\beta_i x}{L} - \sin \frac{\beta_i x}{L}\right)\right] \tag{10.34}$$
$$i = 1, 2...$$

The time dependent function $q(t)$, is now governed by the following equation

$$\ddot{q}(t) + \omega^2 q(t) = 0 \tag{10.35}$$

which indicates that $q(t)$ is harmonic with frequency ω. For the infinite number of β_is, we have an infinite number of corresponding frequencies

$$\omega_i = \frac{\beta_i^2}{L^2} \sqrt{\frac{EI}{\rho}} \tag{10.36}$$

Generally ω_i is called the natural frequency of the mode $q_i(t)$. It follows that an infinite number of solutions exist

$$q_i(t) = B_i \cos \omega_i t + D_i \sin \omega_i t \tag{10.37}$$

where B_i and D_i are constants to be determined from the initial conditions later.

Note that (10.9) is linear and homogeneous, from the Superposition or Linearity Principle, a solution $\omega(x,t)$ can be given by

$$\omega_i(x,t) = \sum_{i=1}^{\infty} \phi_i(x) q_i(t) \tag{10.38}$$

By introducing the following orthogonal conditions

$$\rho \int_0^L \phi_i \phi_j dx + M\phi_i(L)\phi_j(L) = \begin{cases} 0 & \text{if } i \neq j \\ \rho & \text{if } i = j \end{cases}$$

$$EI \int_0^L \phi_i'' \phi_j'' dx = \begin{cases} 0 & \text{if } i \neq j \\ \omega_i^2 & \text{if } i = j \end{cases} \tag{10.39}$$

it can be easily to determine A_i, B_i and D_i

$$A_i = [\frac{1}{\int_0^L \bar{\phi}_i^2(x)dx + \frac{M}{\rho}\bar{\phi}_i^2(L)}]^{\frac{1}{2}} \tag{10.40}$$

$$B_i = \int_0^L \omega(x,0)\phi_i(x)dx + \frac{M}{\rho}\omega(L,0)\phi_i(L) \tag{10.41}$$

$$D_i = \frac{1}{\omega_i}[\int_0^L \dot{\omega}(x,0)\phi_i(x)dx + \frac{M}{\rho}\dot{\omega}(L,0)\phi_i(L)] \tag{10.42}$$

Moreover, from $\int_0^L \bar{\phi}_i^2(x) = L$ when $M = 0$, A_i can be simplified to

$$A_i = \begin{cases} \sqrt{L}/L & \text{when } M = 0 \\ 1/[L + \frac{\rho L^2}{M\beta_i^2}(\frac{1+\cosh\beta_i \cos\beta_i}{\sinh\beta_i \sin\beta_i})^2]^{\frac{1}{2}} & \text{when } M \neq 0 \end{cases}$$

It should be noted that the solution $\omega(x,t)$ obtained above is only valid for the conservative Euler–Bernoulli beam vibration system.

Defining the generalized coordinates vector as

$$Q = [q_1\ q_2\ \dots\ q_N]^T \tag{10.43}$$

we can re-write the kinetic energy E_k into the following compact form

$$E_k = \frac{1}{2}\dot{Q}^T M_A \dot{Q} \tag{10.44}$$

considering the following equation

$$\int_0^L \dot{\omega}^2(x,t)dx = \sum_{i=1}^{N} \dot{q}_i^2(t)[1 - \frac{M}{\rho}\phi_i^2(L)] + 2\sum_{i,j=1\ i\neq j}^{N} \dot{q}_i\dot{q}_j[-\frac{M}{\rho}\phi_i(L)\phi_j(L)] \quad (10.45)$$

where $M_A \in R^{N\times N}$ is the symmetric and positive definite inertia matrix of the system which is given by

$$M_A = \begin{bmatrix} \gamma_A^1 & b_A^{12} & \cdots & b_A^{1N} \\ b_A^{12} & \gamma_A^2 & \cdots & b_A^{2N} \\ \vdots & \vdots & \vdots & \vdots \\ b_A^{1N} & b_A^{2N} & \cdots & \gamma_A^N \end{bmatrix} \quad (10.46)$$

where

$$\gamma_A^i = \rho[1 - \frac{M}{\rho}\phi_i^2(L)] + M\phi_i^2(L) = \rho \quad (10.47)$$

$$b_A^{ij} = 0 \quad (10.48)$$

we can re-write the kinetic energy E_p into the following compact form

$$E_p = \frac{1}{2}\sum_{i=1}^{N} q_i^2\omega_i^2\rho = \frac{1}{2}Q^T K_A Q \quad (10.49)$$

where K_A, the stiffness matrix of the system, is given by

$$K_A = \text{diag}[\omega_1^2\rho\ \omega_2^2\rho\ \ldots\ \omega_N^2\rho] \quad (10.50)$$

By using the Euler–Lagrange's equations

$$M_A\ddot{Q} + K_A Q = \tau \quad (10.51)$$

10.3 Control Design

For the system in this chapter, we designed a neural network control with full state feedback and output feedback in this section, respectively. In addition, we theoretically prove the stability of the controller by Lyapunov's direct method.

In the process of the control design, we adopt the backstepping method which is to decompose the complex nonlinear system into subsystems may not exceed the number of system order. Subsequently, by using the iterative design method, through coordinate transformation, introduce a virtual stabilization function, design the control torque starting from the first subsystem without control input signal, and gradually backstepping to the last subsystem with the control input, and then the real control input signal is obtained.

10.3.1 Full-State Feedback

Transform the Eq. (10.51), we have the description of the system dynamics

$$\ddot{Q} = M_A^{-1}[\tau - K_A Q]. \tag{10.52}$$

Given the errors of system states z_1 and z_2 as follows

$$z_1 = Q - Q_d, \tag{10.53}$$
$$z_2 = \dot{Q} - \mu_1. \tag{10.54}$$

where μ_1 is the virtual control to z_1. We choose $\mu_1 = -K_1 z_1 + \dot{x}_d$, $K_1 = K_1^T > 0$. Then the derivative of the error variables can be described as

$$\dot{z}_1 = z_2 + \mu_1 - \dot{Q}_d = z_2 - K_1 z_1. \tag{10.55}$$
$$\dot{z}_2 = \ddot{Q} - \dot{\mu}_1 = M_A^{-1}[\tau - K_A Q] - \dot{\mu}_1. \tag{10.56}$$

Design V_1 as the initial Lyapunov function

$$V_1 = \frac{1}{2} z_1^T z_1, \tag{10.57}$$

Further, $\dot{V}_1$ can be obtained

$$\dot{V}_1 = z_1^T (z_2 - K_1 z_1) = z_1^T z_2 - z_1^T K_1 z_1. \tag{10.58}$$

Aiming at removing the coupling term $z_1^T z_2$ in (10.58), the Lyapunov function as follows can be employed.

$$V_2 = \frac{1}{2} z_1^T z_1 + \frac{1}{2} z_2^T M_A z_2, \tag{10.59}$$

Further, $\dot{V}_2$ can be obtained

$$\begin{aligned} \dot{V}_2 &= z_1^T \dot{z}_1 + z_2^T M_A \dot{z}_2 \\ &= z_1^T z_2 - z_1^T K_1 z_1 + z_2^T (\tau - K_A Q - M_A \dot{\mu}_1). \end{aligned} \tag{10.60}$$

Then, the model-based controller τ_0 is proposed

$$\tau_0 = -z_1 - K_2 z_2 + K_A Q + M_A \dot{\mu}_1 \tag{10.61}$$

Substituting Eq. (10.61) into (10.60)

$$\dot{V}_2 = -z_1^T K_1 z_1 - z_2^T K_2 z_2 \tag{10.62}$$

To guarantee the closed-loop system stability, $\dot{V}_2$ requires to be negative definite. Therefore, control parameters $K_1 > 0, K_2 > 0$. Since $M_A\dot{\mu}_1$ and $K_A Q$ are uncertain, the control law (10.61) may be infeasible. Therefore, neural network is introduced to deal with the dynamic uncertainty via the online estimation.

$$\tau = -z_1 - K_2 z_2 + \hat{W}^T S(Z) \tag{10.63}$$

where $\hat{W}_i$ is the estimation of neural network weight. $S_i(Z) \in \mathbb{R}^l$ is radial basis function (RBF). $Z = [Q, \dot{Q}, \mu_1, \dot{\mu}_1]^T$ denotes the inputs to the adaptive RBF neural network.

The adaptive law $\dot{\hat{W}}_i$ is proposed

$$\dot{\hat{W}}_i = -\Gamma_i(S_i(Z)z_{2,i} + \sigma_i \hat{W}_i) \tag{10.64}$$

where, because the system needs strong robustness, a small constant $\sigma_i > 0$ is introduced. $\hat{W}_i^T$ is the approximation of neural networks weights W_i^{*T}.

$$W_i^{*T} S_i(Z) = K_A Q + M_A \dot{\mu}_1 - \epsilon(Z) \tag{10.65}$$

Theorem 10.1 *For the dynamic model of the flexible beam, by employing the control input and the adaptive law, because the full state information is known, for each compact set Ω_0 where $(x_1(0), x_2(0), \hat{W}_1(0) \ldots \hat{W}_n(0)) \in \Omega_0$, the initial conditions are bounded and the trajectories of closed-loop beam system are semiglobally uniformly bounded. The error signals z_1, z_2 and $\tilde{W}$ will remain within $\Omega_{z_1}, \Omega_{z_2}, \Omega_W$:*

$$\Omega_{z_1} := \{z_1 \in \mathbb{R}^n \mid \| z_1 \| \leq \sqrt{X}\}, \tag{10.66}$$

$$\Omega_{z_2} := \{z_2 \in \mathbb{R}^n \mid \| z_2 \| \leq \sqrt{\frac{X}{\lambda_{\min}(M_A)}}\}, \tag{10.67}$$

$$\Omega_{\tilde{W}} := \{\tilde{W} \in \mathbb{R}^{l\times n} \mid \| \tilde{W} \| \leq \sqrt{\frac{X}{\lambda_{\min}(\Gamma^{-1})}}\}, \tag{10.68}$$

where $X = 2(V_3(0) + \frac{b_1}{a_1})$, $b_1 > 0$ and $a_1 > 0$.

Proof We propose the Lyapunov function candidate as follows

$$V_3 = V_2 + \frac{1}{2}\sum_{i=1}^{n} \tilde{W}_i^T \Gamma_i^{-1} \tilde{W}_i \tag{10.69}$$

Then, $\dot{V}_3$ can be derived

$$\begin{aligned}\dot{V}_3 &= z_1^T z_2 - z_1^T K_1 z_1 + z_2^T(\tau - K_A Q - M_A\dot{\mu}_1)\\&\quad - \sum_{i=1}^{n} \tilde{W}_i^{\,T}(S_i(Z)z_{2,i} + \sigma_i \hat{W}_i)\\&= -z_1^T K_1 z_1 - z_2^T K_2 z_2 + z_2^T \hat{W}^T S(Z) - z_2^T \epsilon(Z)\\&\quad -z_2^T W^{*T} S(Z) - \sum_{i=1}^{n} \tilde{W}_i^{\,T}(S_i(Z)z_{2,i} + \sigma_i \hat{W}_i) \end{aligned}\tag{10.70}$$

Consider the following inequalities

$$\begin{aligned}-\sum_{i=1}^{n}\sigma_i \tilde{W}_i^{\,T}\hat{W}_i &= -\sum_{i=1}^{n}\sigma_i \tilde{W}_i^{\,T}(W_i^* + \tilde{W}_i)\\&\leq -\frac{1}{2}\sum_{i=1}^{n}\sigma_i(\| \tilde{W}_i \|^2 + \| W_i^* \|^2)\end{aligned}\tag{10.71}$$

$$-z_2^T\epsilon(Z) \leq \frac{1}{2}(\| \bar{\epsilon}(Z) \|^2 + z_2^T z_2)\tag{10.72}$$

We have

$$\begin{aligned}\dot{V}_3 &\leq -z_1^T K_1 z_1 - z_2^T(K_2 - \frac{1}{2}I)z_2 - \sum_{i=1}^{n}\tilde{W}_i^{\,T} S_i(Z) z_{2,i}\\&\quad -\sum_{i=1}^{n}\frac{\sigma_i}{2}\| \tilde{W}_i \|^2 + \sum_{i=1}^{n}\frac{\sigma_i}{2}\| W_i^* \|^2 + \frac{1}{2}\|\bar{\epsilon}\|^2\\&\leq -a_1 V_3 + b_1,\end{aligned}\tag{10.73}$$

where

$$\begin{aligned}a_1 &= \min(2\lambda_{\min}(K_1), \frac{2\lambda_{\min}(K_2 - \frac{1}{2}I_{n\times n})}{\lambda_{\max}(M_A)}, \min_{i=1,\ldots,n}(\frac{\sigma_i}{\lambda_{\max}(\Gamma_i^{-1})})),\\ b_1 &= \sum_{i=1}^{n}\frac{\sigma_i}{2}\| W_i^* \|^2 + \frac{1}{2}\| \bar{\epsilon}(Z) \|^2\end{aligned}\tag{10.74}$$

where $\lambda_{\max}$ and $\lambda_{\min}$ respectively are the maximum and minimum eigenvalues. In order to guarantee $a_1 > 0$, the control parameters should be chosen to satisfy $\lambda_{\min}(K_1) > 0$ and $\lambda_{\min}(K_2 - \frac{1}{2}I) > 0$.

Multiplying (10.73) by $e^{a_1 t}$, and integrating, we have

$$V_3 \leq (V_3(0) - \frac{a_1}{b_1})e^{-a_1 t} + \frac{a_1}{b_1} \leq V_3(0) + \frac{a_1}{b_1}\tag{10.75}$$

Thus

$$\frac{1}{2} \parallel z_1 \parallel^2 \leq V_3(0) + \frac{a_1}{b_1} \tag{10.76}$$

From the analysis above, the system of the flexible beam is stabilized via Lyapunov function that we design.

10.3.2 Output Feedback

However, the control law with full-state feedback (10.61) needs to satisfy that both Q and $\dot{Q}$ can be obtained by measurement. In practice, if $\dot{Q}$ is immesurable, a high-gain observer can be introduced to estimate $\dot{Q}$ by employing separation principle and the certainty equivalence property.

$\frac{\pi_{k+1}}{\varepsilon^k}$ converges asymptotically to $Q^{(k)}$. As a result of the high-gain $\frac{1}{\varepsilon}$, ξ_k converges to zero with a small time constant if Q and its k derivatives are bounded. Therefore, the observer $\frac{\pi_{k+1}}{\varepsilon^k}$ is appropriate to estimate the system unmeasurable states up to the nth order. Therefore,

$$Q = \pi_1, \hat{\dot{Q}} = \frac{\pi_2}{\varepsilon}, \hat{z}_2 = \frac{\pi_2}{\varepsilon} - \alpha_1 \tag{10.77}$$

where

$$\varepsilon\dot{\pi}_1 = \pi_2, \varepsilon\dot{\pi}_2 = -\bar{\lambda}_1\pi_2 - \pi_1 + Q. \tag{10.78}$$

Rewrite the control turque (10.63) and adaptive law (10.64) by backstepping.

$$\tau_2 = -z_1 - K_2\hat{z}_2 + \hat{W}^T S(\hat{Z}), \tag{10.79}$$

$$\dot{\hat{W}}_i = -\Gamma_i(S_i(\hat{Z})\hat{z}_{2,i} + \sigma_i \hat{W}_i), \tag{10.80}$$

Theorem 10.2 *For the dynamic model, by employing the control input and the adaptive law, because the output information is known, for each compact set* Ω_0 *where* $(x_1(0), x_2(0), \hat{W}_1(0) \ldots \hat{W}_n(0)) \in \Omega_0$, *the initial conditions are bounded and the trajectories of closed-loop beam system are semiglobally uniformly bounded. The error signals* z_1, z_2 *and* $\tilde{W}$ *will remain within* $\Omega_{z_1}, \Omega_{z_2}, \Omega_W$*:*

$$\Omega_{z_1} := \{z_1 \in \mathbb{R}^n \mid \parallel z_1 \parallel \leq \sqrt{X}\}, \tag{10.81}$$

$$\Omega_{z_2} := \{z_2 \in \mathbb{R}^n \mid \parallel z_2 \parallel \leq \sqrt{\frac{X}{\lambda_{\min}(M_A)}}\}, \tag{10.82}$$

$$\Omega_{\tilde{W}} := \{\tilde{W} \in \mathbb{R}^{l\times n} \mid \| \tilde{W} \| \leq \sqrt{\frac{X}{\lambda_{\min}(\Gamma^{-1})}}\}, \tag{10.83}$$

where $X = 2(V_4(0) + \frac{b_2}{a_2})$, $b_2 > 0$ *and* $a_2 > 0$.

Proof We propose the Lyapunov function candidate as follows

$$V_4 = \frac{1}{2}z_1^T z_1 + \frac{1}{2}z_2^T M_A z_2 + \frac{1}{2}\sum_{i=1}^{n} \tilde{W}_i^T \Gamma_i^{-1}\tilde{W}_i \tag{10.84}$$

Then, $\dot{V}_4$ can be derived

$$\begin{aligned}\dot{V}_4 \leq & -z_1^T K_1 z_1 - z_2^T(K_2 - \frac{1}{2}I)z_2 - z_2^T K_2 \tilde{z}_2 \\ & + \frac{1}{2} \| \bar{\epsilon} \|^2 + \sum_{i=1}^{n} z_{2,i}[\hat{W}_i^T S_i(\hat{Z}) - W_i^{*T} S_i(Z)] \\ & - \sum_{i=1}^{n}[\tilde{W}_i^T S_i(\hat{Z})\hat{z}_{2,i} + \sigma_i \tilde{W}_i^T \hat{W}_i]\end{aligned} \tag{10.85}$$

where

$$\xi_2 = \frac{\pi_2}{\varepsilon} - \dot{Q} = -\varepsilon\psi^{(2)}, \tag{10.86}$$

$$\tilde{z}_2 = \hat{z}_2 - z_2 = \frac{\pi_2}{\varepsilon} - \mu_1 - \dot{Q} + \alpha_1 = \xi_2. \tag{10.87}$$

where ε is an any small constant, $\psi = \pi_2 + \bar{\lambda}_1\pi_1$. For $\forall t > t^* > 0$ and $h_2 > 0$, then $\| \xi_2 \| \leq \varepsilon h_2$. Therefore, $\dot{Q}$ can be estimated by $\frac{\pi_2}{\varepsilon}$.

$$\hat{\dot{Q}} = \frac{\pi_2}{\varepsilon}, \tag{10.88}$$

$$\hat{z}_2 = \frac{\pi_2}{\varepsilon} - \mu_1 \tag{10.89}$$

Using the following inequality

$$-\sigma_i \tilde{W}_i^T \hat{W}_i \leq -\sigma_i(\frac{\| \tilde{W}_i \|^2}{2} - \frac{\| W_i^* \|^2}{2}), \tag{10.90}$$

$$\| S_i(\hat{Z}) \|^2 \leq l_i, \tag{10.91}$$

$$\begin{aligned}\hat{W}_i^T S_i(\hat{Z}) &= (\tilde{W}_i^T + W_i^{*T})(S_i(Z) + \epsilon S_{ti}) \\ &= W_i^{*T}\epsilon S_{ti} + W_i^{*T} S_i(Z) + \tilde{W}_i^T S_i(Z)\end{aligned} \tag{10.92}$$

where ϵ and l_i are positive and S_{ti} is bounded.

We have

$$\dot{V}_4 \le -z_1^T K_1 z_1 - z_2^T (K_2 - \frac{1}{2}I) z_2 - z_2^T K_2 \tilde{z}_2 + \sum_{i=1}^{n} z_{2,i} W_i^{*T} \epsilon S_{ti} + \frac{1}{2} \parallel \bar{\epsilon} \parallel^2 - \sum_{i=1}^{n} \tilde{W}_i^T S_i(\hat{Z}) \hat{z}_{2,i} + \frac{1}{2} \sum_{i=1}^{n} \sigma_i \parallel W_i^* \parallel^2 - \frac{1}{2} \sum_{i=1}^{n} \sigma_i \parallel \tilde{W}_i \parallel^2 \quad (10.93)$$

Apply the following inequalities

$$\sum_{i=1}^{n} z_{2,i} W_i^{*T} \epsilon S_{ti} \le \frac{1}{2} z_2^T z_2 + \sum_{i=1}^{n} \frac{\parallel W_i^* \parallel^2 \epsilon^2 \parallel S_{ti} \parallel^2}{2} \quad (10.94)$$

$$-\sum_{i=1}^{n} \tilde{W}_i^T S_i(\hat{Z}) \tilde{z}_{2,i} = -\sum_{i=1}^{n} \frac{\sqrt{\sigma_i} \tilde{W}_i^T}{\sqrt{2}} \frac{\sqrt{2} S_i(\hat{Z}) \tilde{z}_{2,i}}{\sqrt{\sigma_i}} \le \sum_{i=1}^{n} \frac{\sigma_i \parallel \tilde{W}_i \parallel^2}{4} + \sum_{i=1}^{n} \frac{2 \parallel S_i(\hat{Z}) \parallel^2}{\sigma_i} \frac{1}{2} \tilde{z}_2^T \tilde{z}_2 \quad (10.95)$$

We can obtain

$$\begin{aligned}
\dot{V}_4 \le & -z_1^T K_1 z_1 - z_2^T (K_2 - \frac{1}{2}I) z_2 - z_2^T K_2 \tilde{z}_2 + \frac{1}{2} z_2^T z_2 \\
& + \sum_{i=1}^{n} \frac{\parallel W_i^* \parallel^2 \epsilon^2 \parallel S_{ti} \parallel^2}{2} + \frac{1}{2} \parallel \bar{\epsilon} \parallel^2 + \sum_{i=1}^{n} \frac{\sigma_i \parallel \tilde{W}_i \parallel^2}{4} \\
& + \sum_{i=1}^{n} \frac{2 \parallel S_i(\hat{Z}) \parallel^2}{\sigma_i} \frac{1}{2} \tilde{z}_2^T \tilde{z}_2 + \frac{1}{2} \sum_{i=1}^{n} \sigma_i \parallel W_i^* \parallel^2 - \frac{1}{2} \sum_{i=1}^{n} \sigma_i \parallel \tilde{W}_i \parallel^2 \\
\le & -z_1^T K_1 z_1 - z_2^T (K_2 - I) z_2 - z_2^T K_2 \tilde{z}_2 + \frac{1}{2} \parallel \bar{\epsilon} \parallel^2 - \sum_{i=1}^{n} \frac{\sigma_i}{4} \parallel \tilde{W}_i \parallel^2 \\
& + \sum_{i=1}^{n} \frac{2 \parallel S_i(\hat{Z}) \parallel^2}{\sigma_i} \frac{1}{2} \tilde{z}_2^T \tilde{z}_2 + \sum_{i=1}^{n} (\frac{\sigma_i}{2} + \frac{\epsilon^2 \parallel S_{ti} \parallel^2}{2}) \parallel W_i^* \parallel^2 \quad (10.96)
\end{aligned}$$

Using $z_2^T K_2 \tilde{z}_2 \le \frac{1}{2} z_2^T z_2 + \frac{1}{2} (K_2 \tilde{z}_2)^T (K_2 \tilde{z}_2)$, we further have

$$\dot{V}_4 \le -z_1^T K_1 z_1 - z_2^T(K_2 - \frac{3}{2}I)z_2 - \frac{1}{2}K_2^T K_2 \xi_2^T \xi_2 + \frac{1}{2} \parallel \bar{\epsilon} \parallel^2 - \sum_{i=1}^{n} \frac{\sigma_i}{4} \parallel \tilde{W}_i \parallel^2$$

$$+ \sum_{i=1}^{n} \frac{2l_i}{\sigma_i} \frac{1}{2} \xi_2^T \xi_2 + \frac{1}{2} \sum_{i=1}^{n} (\sigma_i + \epsilon^2 \parallel S_{ti} \parallel^2) \parallel W_i^* \parallel^2$$

$$\le -z_1^T K_1 z_1 - z_2^T(K_2 - \frac{3}{2}I)z_2 + \frac{1}{2} \parallel \bar{\epsilon} \parallel^2 - \sum_{i=1}^{n} \frac{\sigma_i}{4} \parallel \tilde{W}_i \parallel^2$$

$$+\lambda_{\max}(K_2^T K_2 + \mathrm{diag}[\frac{2l_i}{\sigma_i}]) \frac{1}{2} \xi_2^T \xi_2 + \frac{1}{2} \sum_{i=1}^{n} (\sigma_i + \epsilon^2 \parallel S_{ti} \parallel^2) \parallel W_i^* \parallel^2 \tag{10.97}$$

Applying $\frac{1}{2}\xi_2^T \xi_2 \le \frac{1}{2}\varepsilon^2 h_2^2$, we have

$$\dot{V}_4 \le -z_1^T K_1 z_1 - z_2^T(K_2 - \frac{3}{2}I)z_2 - \sum_{i=1}^{n} \frac{\sigma_i}{4} \parallel \tilde{W}_i \parallel^2$$

$$+\frac{1}{2} \parallel \bar{\epsilon} \parallel^2 + \lambda_{\max}(K_2^T K_2 + \mathrm{diag}[\frac{2l_i}{\sigma_i}]) \frac{1}{2} \varepsilon^2 h_2^2$$

$$+\frac{1}{2} \sum_{i=1}^{n} (\sigma_i + \epsilon^2 \parallel S_{ti} \parallel^2) \parallel W_i^* \parallel^2$$

$$\le -a_2 V_4 + b_2, \tag{10.98}$$

where a_2 and b_2 are two constants defined as

$$a_2 = \min(2\lambda_{\min}(K_1), \frac{2\lambda_{\min}(K_2 - \frac{3}{2}I)}{\lambda_{\max}(M_A)}, \min_{i=1,2\ldots n}(\frac{\sigma_i}{2\lambda_{\max}(\Gamma_i^{-1})})), \tag{10.99}$$

$$b_2 = \frac{1}{2} \sum_{i=1}^{n} (\epsilon^2 \parallel S_{ti} \parallel^2 + \sigma_i) \parallel W_i^* \parallel^2$$

$$+\lambda_{\max}(K_2^T K_2 + \mathrm{diag}[\frac{2l_i}{\sigma_i}]) \frac{1}{2} \varepsilon^2 h_2^2 + \frac{1}{2} \parallel \bar{\epsilon} \parallel^2. \tag{10.100}$$

To guarantee that $a_2 > 0$, the control parameters K_1 and K_2 require to meet two conditions as follows, respectively.

$$\lambda_{\min}(K_1) > 0, \lambda_{\min}(K_2 - \frac{3}{2}I) > 0. \tag{10.101}$$

Multiplying (10.98) by $e^{a_2 t}$, and integrating, we have

$$V_4 \le (V_4(0) - \frac{a_2}{b_2})e^{-a_2 t} + \frac{a_2}{b_2} \le V_4(0) + \frac{a_2}{b_2} \tag{10.102}$$

Table 10.1 Parameters of the flexible system

Parameter	Description	Value	Unit
L	Total length of link	1.0	m
EI	Uniform flexural rigidity	5.0	Nm^2
ρ	Uniform mass per unit length	1.0	kg/m
M_t	Tip mass of the flexible beam	5.0	kg

Thus

$$\frac{1}{2} \parallel z_1 \parallel^2 \leq V_4(0) + \frac{a_2}{b_2} \tag{10.103}$$

From the analysis above, the system of the flexible beam is stabilized via Lyapunov function that we design.

10.4 Numerical Simulations

Simulations of the flexible beam are carried out from four aspects, the simulation results of open-loop system are shown in the first part; the control performance of proportional-derivative (PD) strategy is introduced in the second part; adaptive neural network control for full-state feedback and output feedback is revealed in the third and fourth parts. The system parameters are in Table 10.1.

Simulation Results for Without Control:

The flexible beam system is vibrating without control force, namely, the control torque $u(t) = 0$. The parameters of the flexible manipulator are chosen as $M_t = 5.0\,\mathrm{kg}$, $N = 10$, and the initial status are selected as $w(L,t) = 1$ rad. In Fig. 10.2, we can see that the deflection of the flexible beam system is varying periodically with the amplitude which may cause serious damage to this system.

Simulation Results for PD Control:

The PD control is employed to suppress vibration of the flexible manipulator with $M_t = 5.0\,\mathrm{kg}$, $N = 10$ and then we have $\beta = 0.870$, $\gamma = 1.171$. The PD control law is designed as $u = -K_p(q_{10} - q_{10d}) - K_d \dot{q_{10}}$, where K_p denotes proportional gain, K_d represents differential gain, q_{10} is the tenth system state and q_{10d} is the desired value. Subsequently, we design $\tau \in \mathbb{R}^n = [0, 0, \ldots, u]^T$.

We choose $K_p = 30$, $K_d = 2$, and the corresponding tip trajectory w using PD strategy is revealed in Fig. 10.3. We can get the conclusion that the real elastic vibration of the system under PD controller converges to a small neighborhood of zero within 7s. In order to reduce the elastic vibration within a shorter time, we introduce a neural network controller for the flexible beam.

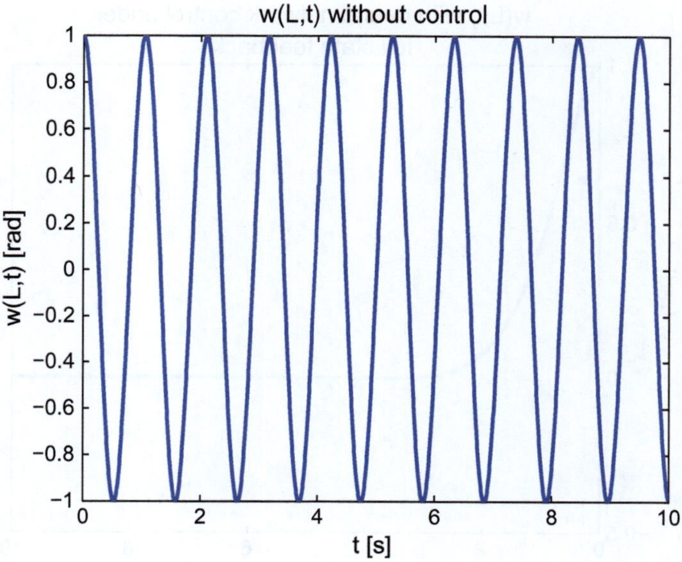

Fig. 10.2 $w(L, t)$ without control

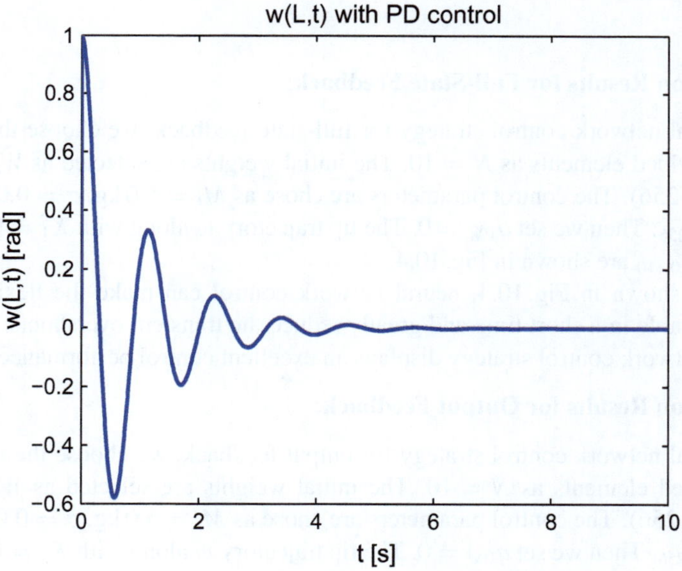

Fig. 10.3 $w(L, t)$ with PD control

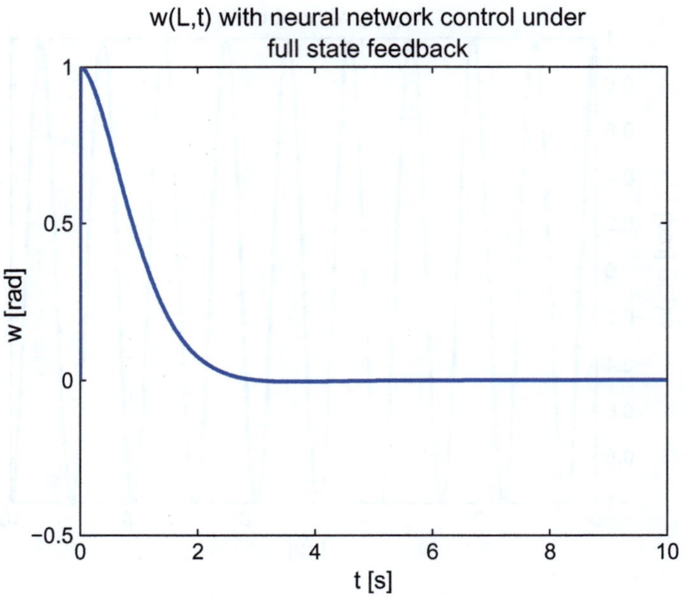

Fig. 10.4 $w(L,t)$ with neural network control under full state feedback

Simulation Results for Full-State Feedback:

For neural network control strategy for full-state feedback, we choose the number of the divided elements as $N = 10$. The initial weights are selected as $\hat{W}_i = 0(i = 1, 2, \ldots, 256)$. The control parameters are chose as $M_t = 5.0\,\text{kg}$, $\sigma = 0.0002$, $\Gamma = 100I_{256\times256}$. Then we set $q_{10d} = 0$. The tip trajectory w along with $K_1 = 0.2I_{10\times10}$, $K_2 = 3I_{10\times10}$ are shown in Fig. 10.4.

As is shown in Fig. 10.4, neural network control can make the flexible beam system stable in a short time and greatly reduce the transient overshoots. Thus, the neural network control strategy displays an excellent control performance.

Simulation Results for Output Feedback:

For neural network control strategy for output feedback, we choose the number of the divided elements as $N = 10$. The initial weights are selected as $\hat{W}_i = 0(i = 1, 2, \ldots, 256)$. The control parameters are chose as $M_t = 5.0\,\text{kg}$, $\sigma = 0.0002$, $\Gamma = 100I_{256\times256}$. Then we set $q_{10d} = 0$. The tip trajectory w along with $K_1 = 0.2I_{10\times10}$, $K_2 = 3I_{10\times10}$ are shown in Fig. 10.5.

As is shown in Fig. 10.5, neural network control can make the flexible beam system stable in a short time and greatly reduce the transient overshoots. Thus, the neural network control strategy displays an excellent control performance.

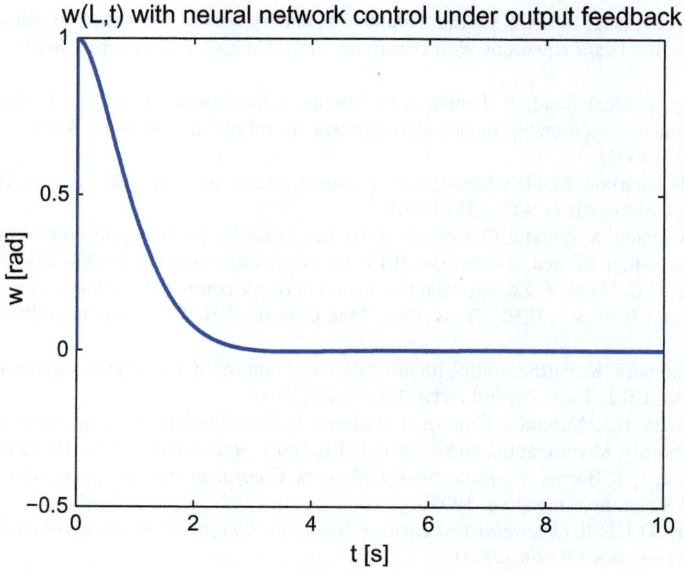

Fig. 10.5 $w(L, t)$ with neural network control under output feedback

10.5 Conclusion

In this chapter, the dynamic model of a flexible beam developed with AMM is investigated. Subsequently, we obtain a discretized model for the flexible beam system. Base on the dynamic model, neural network control are studied to achieve the control objectives. Aiming at ensuring stability rigorously, the closed-loop system is uniform ultimate boundedness (UUB) via the Lyapunov stability. By appropriately choosing design parameters, the state of the system is proven to converge to zero with a small neighborhood. Simulations for the flexible beam system are given to verify the effectiveness of the proposed neural network control strategy, and the final performance is compared with that of PD control strategy.

References

1. S.S. Ge, T.H. Lee, G. Zhu, Energy-based robust controller design for multi-link flexible robots. Mechatronics **6**(7), 779–798 (1996)
2. Y. Li, G. Liu, T. Hong, K. Liu, Robust control of a two-link flexible manipulator with neural networks based quasi-static deflection compensation, in *2003 Proceedings of the American Control Conference*, vol. 6 (IEEE, 2003), pp. 5258–5263
3. S.K. Pradhan, B. Subudhi, Real-time adaptive control of a flexible manipulator using reinforcement learning. IEEE Trans. Autom. Sci. Eng. **9**(2), 237–249 (2012)
4. L. Tian, C. Collins, Adaptive neuro-fuzzy control of a flexible manipulator. Mechatronics **15**(10), 1305–1320 (2005)

5. W. He, S. Zhang, S.S. Ge, C. Liu, Adaptive boundary control for a class of inhomogeneous Timoshenko beam equations with constraints. IET Control Theory Appl. **8**(14), 1285–1292 (2014)
6. J. Weng, J. Mcclelland, A. Pentland, O. Sporns, I. Stockman, M. Sur, E. Thelen, Artificial intelligence. Autonomous mental development by robots and animals. Science **291**(5504), 599–600 (2001)
7. M.M. Polycarpou, Stable adaptive neural control scheme for nonlinear systems. IEEE Trans. Autom. Control **41**(3), 447–451 (1996)
8. Y. Li, S. Qiang, X. Zhuang, O. Kaynak, Robust and adaptive backstepping control for nonlinear systems using RBF neural networks. IEEE Trans. Neural Netw. **15**(3), 693–701 (2004)
9. S.S. Ge, C.C. Hang, T. Zhang, Adaptive neural network control of nonlinear systems by state and output feedback. IEEE Trans. Syst. Man Cybern. Part B (Cybernetics) **29**(6), 818–828 (1999)
10. K.S. Narendra, K. Parthasarathy, Identification and control of dynamical systems using neural networks. IEEE Trans. Neural Netw. **1**(1), 4–27 (1990)
11. A.U. Levin, K.S. Narendra, Control of nonlinear dynamical systems using neural networks. ii. Observability, identification, and control. IEEE Trans. Neural Netw. **7**(1), 30–42 (1996)
12. T.H. Lee, C.J. Harris, *Adaptive Neural Network Control of Robotic Manipulators*, vol. 19 (World Scientific, Singapore, 1998)
13. C. Wang, D.J. Hill, *Deterministic Learning Theory for Identification, Recognition, and Control* (CRC Press, Boca Raton, 2009)
14. Y.J. Liu, C.L. Chen, G.X. Wen, S. Tong, Adaptive neural output feedback tracking control for a class of uncertain discrete-time nonlinear systems. IEEE Trans. Neural Netw. **22**(7), 1162–7 (2011)
15. S.L. Dai, C. Wang, M. Wang, Dynamic learning from adaptive neural network control of a class of nonaffine nonlinear systems. IEEE Trans. Neural Netw. Learn. Syst. **25**(1), 111 (2014)
16. S.L. Dai, C. Wang, F. Luo, Identification and learning control of ocean surface ship using neural networks. IEEE Trans. Ind. Inform. **8**(34), 801–810 (2012)
17. W. He, S.S. Ge, B.V.E. How, Y.S. Choo, *Dynamics and Control of Mechanical Systems in Offshore Engineering* (Springer, London, 2014)
18. R.R. Selmic, F.L. Lewis, Deadzone compensation in motion control systems using neural networks. IEEE Trans. Autom. Control **45**(4), 602–613 (1998)
19. F. Sun, Z. Sun, Y. Zhu, W. Lu, Stable neuro-adaptive control for robots with the upper bound estimation on the neural approximation errors. J. Intell. Robot. Syst. **26**(1), 91–100 (1999)
20. R. Cui, B. Ren, S.S. Ge, Synchronised tracking control of multi-agent system with high-order dynamics. IET Control Theory Appl. **6**(5), 603–614 (2012)
21. K.S. Narendra, S. Mukhopadhyay, Adaptive control using neural networks and approximate models. IEEE Trans. Neural Netw. **8**(3), 475–85 (1997)
22. M. Kawato, K. Furukawa, R. Suzuki, A hierarchical neural-network model for control and learning of voluntary movement. Biol. Cybern. **57**(3), 169–185 (1987)
23. Z. Xiong, Z. Jie, A batch-to-batch iterative optimal control strategy based on recurrent neural network models. J. Process Control **15**(1), 11–21 (2005)
24. C. Guoping, H. Jiazhen, Assumed mode method of a rotating flexible beam. Acta Mechanica Sinica **37**(1), 48–56 (2005)
25. D.S. Cho, N. Vladimir, T.M. Choi, Approximate natural vibration analysis of rectangular plates with openings using assumed mode method. Int. J. Nav. Archit. Ocean Eng. **5**(3), 478–491 (2013)
26. A. Yan, Frequency analysis of a rotating cantilever beam using assumed mode method with coupling effect. Mech. Based Design Struct. Mach. **34**(1), 25–47 (2006)

Chapter 11
Vibration Control of a Three-Dimensional Flexible Beam

11.1 Introduction

In the previous works, the vibration of the system is only considered in one-dimensional space. However, the coupling between the axial deformation and transverse displacement could deteriorate the performance of the controlled system and even lead to the instability. Therefore, several works have been done in modeling and control design with taking the coupling effect into account. In [1], an active boundary control scheme of an axially moving string is proposed to suppress both longitudinal and transverse vibrations and track the desired speed. In [2], boundary control and adaptive control for a two-dimensional string system are developed. In [3], both the axial deformation and the transverse displacement of a marine riser are considered and boundary control is achieved by two actuators in transverse and longitudinal directions to reduce the riser's vibrations. In [4], boundary controller for two-dimensional marine risers modeled as an Euler–Bernoulli beam with only bending couplings is designed. And in [5], the coupling between bending-bending and longitudinal-bending is included in the mathematical modeling. However, in this reference, coupling effects and boundary payloads are not considered together in modeling and control design.

In this chapter, we are going to design boundary control based on the PDE model applying at the top boundary of the beam with three actuators to reduce vibrations in X, Y and Z directions, respectively. The control input consists of the feedback signal of displacement, the time derivative of the displacement and the time derivative of the slope, which can be obtained in practice.

The rest of this chapter is organized as follows. The dynamic model is derived using Hamilton principle in Sect. 11.2. Control design including model-based boundary control and adaptive boundary control designed to compensate the uncertainty of tip payload is proposed in Sect. 11.3. To illustrate the effectiveness of the controller, numerical simulations are presented in Sect. 11.4. Finally, we conclude this chapter in Sect. 11.5.

W. He and J. Liu, *Active Vibration Control and Stability Analysis of Flexible Beam Systems*, https://doi.org/10.1007/978-981-10-7539-1_11

11.2 Problem Formulation

Figure 11.1 is a three-dimensional Euler–Bernoulli beam with its end fixed on at origin and tip clamped with a tip payload. The control is implemented at the tip by the actuator. In this chapter, the effect of gravity is ignored. For clarify, the notations $(\dot{*}) = \frac{\partial(*)}{\partial t}$, $(\ddot{*}) = \frac{\partial^2(*)}{\partial t^2}$, $(*)' = \frac{\partial(*)}{\partial s}$, $(*)'' = \frac{\partial^2(*)}{\partial s^2}$, $(*)'''' = \frac{\partial^4(*)}{\partial s^4}$, $(\dot{*})' = \frac{\partial^2(*)}{\partial t \partial s}$, $(\dot{*})'' = \frac{\partial^3(*)}{\partial t \partial s^2}$, $(*)_0 = *(0,t)$, $(\dot{*})_0 = \frac{\partial(*(0,t))}{\partial t}$, $(\ddot{*})_0 = \frac{\partial^2(*(0,t))}{\partial t^2}$, $(*)'_0 = \frac{\partial(*(0,t))}{\partial s}$, $(*)''_0 = \frac{\partial^2(*(0,t))}{\partial s^2}$, $(*)'''_0 = \frac{\partial^3(*(0,t))}{\partial s^3}$, $(\dot{*})_L = \frac{\partial(*(L,t))}{\partial t}$, $(\ddot{*})_L = \frac{\partial^2(*(L,t))}{\partial t^2}$, $(*)'_L = \frac{\partial(*(L,t))}{\partial s}$, $(*)''_L = \frac{\partial^2(*(L,t))}{\partial s^2}$, $(*)'''_L = \frac{\partial^3(*(L,t))}{\partial s^3}$ are used throughout this chapter.

The kinetic energy of the riser system $E_k(t)$ can be written as

$$E_k(t) = \frac{1}{2}M[(\dot{x}_L)^2 + (\dot{y}_L)^2 + (\dot{z}_L)^2] + \frac{1}{2}\rho\int_0^L [(\dot{x})^2 + (\dot{y})^2 + (\dot{z})^2]ds \tag{11.1}$$

where s and t represent the independent spatial and time variable respectively.

The potential energy $E_p(t)$ due to the tension T, the bending and the axial deformation can be represented as

$$E_p(t) = \frac{1}{2}T\int_0^L [(x')^2 + (y')^2]ds + \frac{1}{2}EI\int_0^L [(x'')^2 + (y'')^2]ds + \frac{1}{2}EA\int_0^L [z' + \frac{1}{2}(x')^2 + \frac{1}{2}(y')^2]^2 ds \tag{11.2}$$

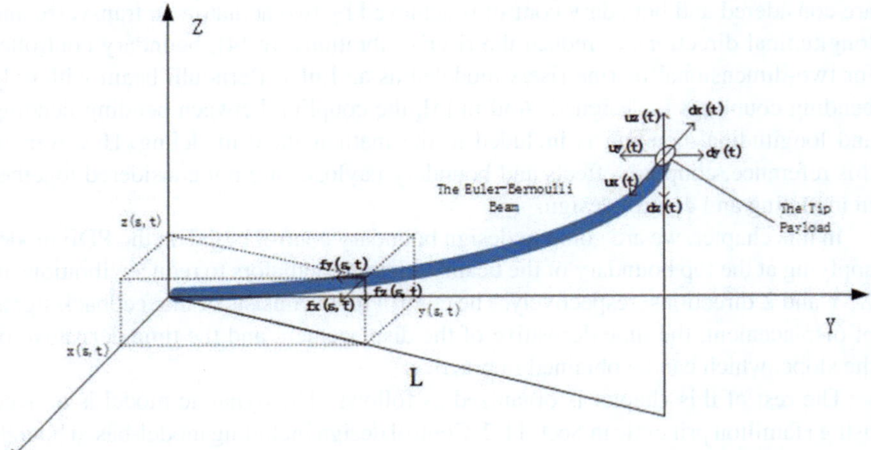

Fig. 11.1 A nonlinear three-dimensional Euler–Bernoulli beam

The virtual work done by unknown distributed external loads $f_x(s,t)$, $f_y(s,t)$ and $f_z(s,t)$ on the riser and time-varying loads $d_x(t)$, $d_y(t)$ and $d_z(t)$ on the tip payload is given by

$$\delta W_f(t) = \int_0^L (f_x\delta x + f_y\delta y + f_z\delta z)ds + d_x\delta x_L + d_y\delta y_L + d_z\delta z_L \tag{11.3}$$

The virtual work done by the control force $u_x(t)$, $u_y(t)$, $u_z(t)$ can be represented as

$$\delta W_m(t) = u_x\delta x_L + u_y\delta y_L + u_z\delta z_L \tag{11.4}$$

Therefore,the virtual work δW done on the system can be represented as

$$\delta W(t) = \delta W_f(t) + \delta W_m(t) \tag{11.5}$$

Using the Hamilton principle $\int_{t_1}^{t_2} \delta[E_k - E_p + W]dt = 0$, we can obtain the following governing equations

$$\rho\ddot{x} = Tx'' + EA(z''x' + x''z') + \frac{3}{2}EA(x')^2x'' + \frac{1}{2}EA[x''(y')^2 + 2x'y'y''] - EIx'''' + f_x \tag{11.6}$$

$$\rho\ddot{y} = Ty'' + EA(z''y' + y''z') + \frac{3}{2}EA(y')^2y'' + \frac{1}{2}EA[y''(x')^2 + 2y'x'x''] - EIy'''' + f_y \tag{11.7}$$

$$\rho\ddot{z} = EAz'' + EAx'x'' + EAy'y'' + f_z \tag{11.8}$$

$\forall(s,t) \in [0, L] \times [0, \infty)$, and the boundary conditions as follows

$$x_0 = y_0 = z_0 = 0 \tag{11.9}$$

$$x_0'' = y_0'' = z_0'' = 0 \tag{11.10}$$

$$x_L'' = y_L'' = z_L'' = 0 \tag{11.11}$$

$$u_x + d_x = M\ddot{x}_L + Tx_L' + \frac{1}{2}EA(x_L')^3 + EAx_L'z_L' + \frac{1}{2}EAx_L'(y_L')^2 - EIx_L''' \tag{11.12}$$

$$u_y + d_y = M\ddot{y}_L + Ty_L' + \frac{1}{2}EA(y_L')^3 + EAy_L'z_L' + \frac{1}{2}EAy_L'(x_L')^2 - EIy_L''' \tag{11.13}$$

$$u_z + d_z = M\ddot{z}_L + EAz_L' + \frac{1}{2}EA(x_L')^2 + \frac{1}{2}EA(y_L')^2 \tag{11.14}$$

$\forall t \in [0, \infty)$.

Assumption 11.1 For the unknown distributed disturbances $f_x(s,t)$, $f_y(s,t)$, $f_z(s,t)$ and unknown boundary disturbances $d_x(t), d_y(t), d_z(t)$, we assume that there exist constants $\bar{f}_x$, $\bar{f}_y$, $\bar{f}_z$, $\bar{d}_x$, $\bar{d}_y$, $\bar{d}_z \in R^+$, such that $f_x(s,t) \le \bar{f}_x$, $f_y(s,t) \le \bar{f}_y$, $f_z(s,t) \le \bar{f}_z$, $\forall[s,t] \in [0,L] \times [0,\infty)$ and $d_x(t) \le \bar{d}_x, d_y(t) \le \bar{d}_y, d_z(t) \le \bar{d}_z$, $\forall t \in [0,\infty)$.

Remark 11.1 This is a reasonable assumption as the disturbances $f_x(s,t)$, $f_y(s,t)$, $f_z(s,t), d_x(t), d_y(t)$ and $d_z(t)$ have finite energy and hence are bounded, i.e. $f_x(s,t)$, $f_y(s,t)$, $f_z(s,t)$, $d_x(t)$, $d_y(t)$, $d_z(t) \in \mathcal{L}_\infty$. The knowledge of exact values of the disturbances are not required.

11.3 Control Design

The control objective is to suppress the vibration of the beam in three-dimensional space, which is to have the transverse and longitudinal displacements $x(s,t)$, $y(s,t)$ and $z(s,t)$ converged to a finite set with the environmental disturbances $f_x(s,t)$, $f_y(s,t)$, $f_z(s,t)$ and $d_x(t)$, $d_y(t)$, $d_z(t)$. In this section, we proposed two cases for the system: (i) model-based boundary control with a certain tip payload, i.e., the tip payload M is known; (ii) adaptive boundary control to compensate the uncertainty of tip payload, i.e., the tip payload M is unknown. Lyapunov's direct method is used to construct both kinds of control inputs u_x, u_y, u_z and analyze the stability of the closed-loop system.

11.3.1 Model-Based Boundary Control with a Certain Tip Payload

Consider the Lyapunov candidate function as

$$V(t) = V_1(t) + V_2(t) + V_3(t), \tag{11.15}$$

where

$$V_1(t) = \frac{1}{2}\rho\int_0^L [(\dot{x})^2 + (\dot{y})^2 + (\dot{z})^2]ds + \frac{1}{2}EA\int_0^L [z' + \frac{1}{2}(x')^2 + \frac{1}{2}(y')^2]^2 ds$$
$$+\frac{1}{2}EI\int_0^L [(x'')^2 + (y'')^2]ds + \frac{1}{2}T\int_0^L [(x')^2 + (y')^2]ds \tag{11.16}$$

$$V_2(t) = \frac{1}{2}M(\dot{x}_L + \beta L x_L')^2 + \frac{1}{2}M(\dot{y}_L + \beta L y_L')^2 + \frac{1}{2}M(\dot{z}_L + \beta L z_L')^2 \tag{11.17}$$

$$V_3(t) = \beta\rho\int_0^L s(\dot{x}x' + \dot{y}y' + \dot{z}z')ds \tag{11.18}$$

where β is a positive constant.

The control laws are given as

$$u_x = -M\beta L\dot{x}'_L - \text{sgn}(\dot{x}_L + \beta L x'_L)\bar{d}_x - 2k_1\dot{x}_L \tag{11.19}$$
$$u_y = -M\beta L\dot{y}'_L - \text{sgn}(\dot{y}_L + \beta L y'_L)\bar{d}_y - 2k_2\dot{y}_L \tag{11.20}$$
$$u_z = -M\beta L\dot{z}'_L - \text{sgn}(\dot{z}_L + \beta L z'_L)\bar{d}_z - 2k_3\dot{z}_L \tag{11.21}$$

where k_1, k_2, k_3 are the control gains.

Lemma 11.1 *The Lyapunov candidate function can be upper and lower bounded as*

$$\lambda_1(\zeta(t) + V_2(t)) \le V(t) \le \lambda_2(\zeta(t) + V_2(t)) \tag{11.22}$$

where $\lambda_1 = \xi_1$ and $\lambda_2 = \xi_2$ are two positive constants and

$$\zeta(t) = \int_0^L [(\dot{x})^2 + (x')^2 + (\dot{y})^2 + (y')^2 + (\dot{z})^2 + (z')^2 + (x')^4 + (y')^4 + (x'y')^2 + (x'')^2 + (y'')^2]ds \tag{11.23}$$

Proof Rewritten $V_1(t)$ as

$$V_1(t) = \frac{1}{2}\rho\int_0^L [(\dot{x})^2 + (\dot{y})^2 + (\dot{z})^2]ds + \frac{1}{2}T\int_0^L [(x')^2 + (y')^2]ds + \frac{1}{2}EA\int_0^L (z')^2 ds + \frac{1}{2}\int_0^L \{z'[(x')^2 + (y')^2] + \frac{1}{4}[(x')^2 + (y')^2]^2\}ds \tag{11.24}$$

we have

$$0 \le \eta_1\zeta(t) \le V_1(t) \le \eta_2\zeta(t) \tag{11.25}$$

where η_1 and η_2 are defined as

$$\eta_1 = \frac{1}{2}\min[\rho, T, EA, EI] \tag{11.26}$$
$$\eta_2 = \frac{1}{2}\max[\rho, T, EA, EI] \tag{11.27}$$

Similarly, we can obtain

$$|V_3(t)| \le \beta\rho L\int_0^L [(\dot{x})^2 + (x')^2 + (\dot{y})^2 + (y')^2 + (\dot{z})^2 + (z')^2]ds \le \beta\rho L\zeta(t) \tag{11.28}$$

Let β satisfy $\beta\rho L < \eta_1$, Let $\xi_1 = \eta_1 - \beta\rho L$, $\xi_2 = \eta_2 + \beta\rho L$, we further have

$$0 \leq \xi_1 V_1(t) \leq V_1(t) + V_3(t) \leq \xi_2 V_1(t). \tag{11.29}$$

Therefore, we have

$$0 \leq \lambda_1[(\zeta(t) + V_2(t)) \leq V(t) \leq \lambda_2(\zeta(t) + V_2(t)), \tag{11.30}$$

where $\lambda_1 = \min(\xi_1, 1)$ and $\lambda_2 = \max(\xi_2, 1)$ are two positive constants.

Lemma 11.2 *The time derivative of the Lyapunov candidate function is upper bounded as*

$$\dot{V}(t) \leq -\lambda V(t) + \varepsilon, \tag{11.31}$$

where λ and ε are two positive constants.

Proof Differentiating V_1 by parts, we obtain

$$\dot{V}_1(t) \leq \delta_1 \int_0^L f_x^2 ds + \delta_2 \int_0^L f_y^2 ds + \delta_3 \int_0^L f_z^2 ds + \frac{1}{\delta_1} \int_0^L (\dot{x})^2 ds + \frac{1}{\delta_2} \int_0^L (\dot{y})^2 ds + \frac{1}{\delta_3} \int_0^L (\dot{z})^2 ds + \Delta_1 \tag{11.32}$$

where Δ_1 is the sum of the boundary term in $\dot{V}_1$. Substituting boundary conditions to Δ_1, we obtain

$$\Delta_1 = (u_x + d_x - M\ddot{x}_L)\dot{x}_L + (u_y + d_y - M\ddot{y}_L)\dot{y}_L + (u_z + d_z - M\ddot{z}_L)\dot{z}_L \tag{11.33}$$

Time derivative of V_2 yields

$$\begin{aligned}\dot{V}_2(t) =& (M\ddot{x}_L + M\beta L\dot{x}'_L)(\dot{x}_L + \beta L x'_L) + (M\ddot{y}_L + M\beta L\dot{y}'_L)(\dot{y}_L + \beta L y'_L) \\ &+ (M\ddot{z}_L + M\beta L\dot{z}'_L)(\dot{z}_L + \beta L z'_L)\end{aligned} \tag{11.34}$$

Time derivative of V_3 yields

$$\dot{V}_3(t) = \beta\rho \int_0^L s(\ddot{x}x' + \dot{x}\dot{x}' + \ddot{y}y' + \dot{y}\dot{y}' + \ddot{z}z' + \dot{z}\dot{z}')ds \tag{11.35}$$

Substituting governing equations, we obtain

$$\begin{aligned}\dot{V}_3(t) =& -\frac{1}{2}\beta\rho \int_0^L (\dot{x})^2 ds - \frac{1}{2}\beta\rho \int_0^L (\dot{y})^2 ds - \frac{1}{2}\beta\rho \int_0^L (\dot{z})^2 ds - \frac{1}{2}\beta EA \int_0^L (z')^2 ds \\ &- \frac{3}{8}\beta EA \int_0^L (x')^4 ds - \frac{3}{8}\beta EA \int_0^L (y')^4 ds - \beta EA \int_0^L (x')^2 z' ds - \beta EA \int_0^L (y')^2 z' ds \\ &- \frac{3}{4}\beta EA \int_0^L (x'y')^2 ds - \frac{1}{2}\beta EI \int_0^L (x'')^2 ds - \frac{1}{2}\beta EI \int_0^L (y'')^2 ds - \frac{1}{2}\beta T \int_0^L (x')^2 ds \\ &- \frac{1}{2}\beta T \int_0^L (y')^2 ds + \beta \int_0^L s(f_x x' + f_y y' + f_z z')ds + \Delta_2\end{aligned} \tag{11.36}$$

where Δ_2 is the sum of the boundary term in $\dot{V}_3$. Substituting the boundary conditions to Δ_2 yields

$$\begin{aligned}\Delta_2 =&\frac{1}{2}\beta\rho L(\dot{x}_L)^2 + \frac{1}{2}\beta\rho L(\dot{y}_L)^2 + \frac{1}{2}\beta\rho L(\dot{z}_L)^2 \\ &+ \beta L[(u_x + d_x - M\ddot{x}_L)x'_L + (u_y + d_y - M\ddot{y}_L)y'_L + (u_z + d_z - M\ddot{z}_L)z'_L] \\ &- \frac{1}{2}\beta TL(x'_L)^2 - \frac{1}{2}\beta TL(y'_L)^2 - \frac{1}{2}\beta EAL\Big\{(z'_L)^2 + z'_L[(x'_L)^2 + (y'_L)^2] \\ &+\frac{1}{4}[(x'_L)^2 + (y'_L)^2]^2\Big\}\end{aligned} \tag{11.37}$$

we have

$$\begin{aligned}\dot{V}_3(t) \le &-\frac{1}{2}\beta\rho\int_0^L(\dot{x})^2ds - \frac{1}{2}\beta\rho\int_0^L(\dot{y})^2ds - \frac{1}{2}\beta\rho\int_0^L(\dot{z})^2ds - \Big(\frac{1}{2}\beta EA - \beta L\delta_6\Big)\int_0^L(z')^2ds \\ &- \frac{3}{8}\beta EA\int_0^L(x')^4ds - \frac{3}{8}\beta EA\int_0^L(y')^4ds - \beta EA\int_0^L(x')^2z'ds - \beta EA\int_0^L(y')^2z'ds \\ &- \frac{3}{4}\beta EA\int_0^L(x'y')^2ds - \frac{1}{2}\beta EI\int_0^L(x'')^2ds - \frac{1}{2}\beta EI\int_0^L(y'')^2ds \\ &- \Big(\frac{1}{2}\beta T - \beta L\delta_5\Big)\int_0^L(y')^2ds + \beta L\int_0^L\Big(\frac{f_x}{\delta_4} + \frac{f_y}{\delta_5} + \frac{f_z}{\delta_6}\Big)ds + \Delta_2 \\ &- \Big(\frac{1}{2}\beta T - \beta L\delta_4\Big)\int_0^L(x')^2ds\end{aligned} \tag{11.38}$$

Δ is the sum of the boundary terms in $\dot{V}$, and applying the control laws and the property of the system $z'_L[(x'_L)^2 + (y'_L)^2] + \frac{1}{4}[(x'_L)^2 + (y'_L)^2]^2 \ge 0$ in Δ, we have

$$\begin{aligned}\Delta \le &- k_1(\dot{x}_L + \beta Lx'_L)^2 - k_2(\dot{y}_L + \beta Ly'_L)^2 - k_3(\dot{z}_L + \beta Lz'_L)^2 \\ &- \Big(k_1 - \frac{1}{2}\beta\rho L\Big)(\dot{x}_L)^2 - \Big(k_2 - \frac{1}{2}\beta\rho L\Big)(\dot{y}_L)^2 - \Big(k_3 - \frac{1}{2}\beta\rho L\Big)(\dot{z}_L)^2 \\ &- \Big(\frac{1}{2}\beta TL - k_1\beta^2L^2\Big)(x'_L)^2 - \Big(\frac{1}{2}\beta TL - k_2\beta^2L^2\Big)(y'_L)^2 \\ &\Big[-\frac{1}{2}\beta EAL - k_3\beta^2L^2)\Big](z'_L)^2\end{aligned} \tag{11.39}$$

Combining $\dot{V}_1(t)$, $\dot{V}_3(t)$ and Δ, we can further obtain

$$\dot{V}(t) \le -\lambda_3[\zeta(t) + V_2(t)] + \varepsilon \tag{11.40}$$

where

$$\lambda_3 = \min\left\{\frac{1}{2}\beta\rho - \frac{1}{\delta_1}, \frac{1}{2}\beta\rho - \frac{1}{\delta_2}, \frac{1}{2}\beta\rho - \frac{1}{\delta_3}, \frac{1}{2}\beta EA - \beta L\delta_6, \frac{3\beta}{8},\right.$$
$$\left.\frac{1}{2}\beta T - \beta L\delta_4, \frac{1}{2}\beta T - \beta L\delta_5, \frac{2k_1}{M}, \frac{2k_2}{M}, \frac{2k_3}{M}\right\} \tag{11.41}$$
$$\varepsilon \leq \left(\alpha\delta_1 + \frac{\beta L}{\delta_4}\right)\bar{f}_x + \left(\alpha\delta_2 + \frac{\beta L}{\delta_5}\right)\bar{f}_x + \left(\alpha\delta_3 + \frac{\beta L}{\delta_6}\right)\bar{f}_x \in \mathcal{L}_\infty \tag{11.42}$$

The design constants are selected to satisfy the following conditions

$$\frac{1}{2}\beta\rho - \frac{1}{\delta_1} \geq 0 \tag{11.43}$$
$$\frac{1}{2}\beta\rho - \frac{1}{\delta_2} \geq 0 \tag{11.44}$$
$$\frac{1}{2}\beta\rho - \frac{1}{\delta_3} \geq 0 \tag{11.45}$$
$$\frac{1}{2}\beta EA - \beta L\delta_6 \geq 0 \tag{11.46}$$
$$\frac{1}{2}\beta T - \beta L\delta_4 \geq 0 \tag{11.47}$$
$$\frac{1}{2}\beta T - \beta L\delta_5 \geq 0 \tag{11.48}$$
$$k_1 - \frac{1}{2}\beta\rho L \geq 0 \tag{11.49}$$
$$k_2 - \frac{1}{2}\beta\rho L \geq 0 \tag{11.50}$$
$$k_3 - \frac{1}{2}\beta\rho L \geq 0 \tag{11.51}$$
$$\frac{1}{2}\beta TL - k_1\beta^2L^2 \geq 0 \tag{11.52}$$
$$\frac{1}{2}\beta TL - k_2\beta^2L^2 \geq 0 \tag{11.53}$$
$$\frac{1}{2}\beta EAL - k_3\beta^2L^2 \geq 0 \tag{11.54}$$

we have

$$\dot{V}(t) \leq -\lambda V(t) + \varepsilon, \tag{11.55}$$

where $\lambda = \lambda_3/\lambda_2$ and $\varepsilon > 0$.

Theorem 11.1 *For the system, under Assumption 11.1, and the control laws, given that the initial conditions are bounded, we can conclude that*

(i) uniform boundedness (UB): the state of the system $x(s,t)$, $y(s,t)$ *and* $z(s,t)$ *will remain in the compact set* Ω_1 *defined by* $\Omega_1 := \{x(s,t), y(s,t), z(s,t) \in R|\ |x(s,t)|, |y(s,t)|, |z(s,t)| \leq D_1, \forall (s,t) \in [0,L] \times [0,\infty)\}$, *where the constant* $D_1 = \sqrt{\frac{L}{\lambda_1}\left(V(0) + \frac{\varepsilon}{\lambda}\right)}$.

(ii) uniform ultimate boundedness (UUB): the state of the system $x(s,t)$, $y(s,t)$ *and* $z(s,t)$ *will eventually converge to the compact* Ω_2 *defined by* $\Omega_2 := \{x(s,t), y(s,t), z(s,t) \in R|\ \lim_{t\to\infty} |x(s,t)|, \lim_{t\to\infty}, \ |y(s,t)|, \lim_{t\to\infty} |z(s,t)| \leq D_2, \forall t \in [0,\infty)\}$, *where the constant* $D_2 = \sqrt{\frac{L\varepsilon}{\lambda_1 \lambda}}$.

Proof

$$\frac{d}{dt}(Ve^{\lambda t}) \leq \varepsilon e^{\lambda t} \tag{11.56}$$

Integration of the above inequality, we obtain

$$V(t) \leq \left(V(0) - \frac{\varepsilon}{\lambda}\right)e^{-\lambda t} + \frac{\varepsilon}{\lambda} \leq V(0)e^{-\lambda t} + \frac{\varepsilon}{\lambda} \in \mathcal{L}_\infty \tag{11.57}$$

which implies $V(t)$ is bounded.

$$\frac{1}{L}[x(s,t)]^2 \leq \int_0^L [x'(s,t)]^2 ds \leq \zeta(t) \leq \frac{1}{\lambda_1} V(t) \in \mathcal{L}_\infty \tag{11.58}$$

$$\frac{1}{L}[y(s,t)]^2 \leq \int_0^L [y'(s,t)]^2 ds \leq \zeta(t) \leq \frac{1}{\lambda_1} V(t) \in \mathcal{L}_\infty \tag{11.59}$$

$$\frac{1}{L}[z(s,t)]^2 \leq \int_0^L [z'(s,t)]^2 ds \leq \zeta(t) \leq \frac{1}{\lambda_1} V(t) \in \mathcal{L}_\infty \tag{11.60}$$

Rearranging the terms of the above three inequalities, we can obtain

$$\begin{aligned} |x(s,t)|, |y(s,t)|, |z(s,t)| &\leq \sqrt{\frac{L}{\lambda_1}\left(V(0)e^{-\lambda t} + \frac{\varepsilon}{\lambda}\right)} \\ &\leq \sqrt{\frac{L}{\lambda_1}\left(V(0) + \frac{\varepsilon}{\lambda}\right)}, \forall (s,t) \in [0,L] \times [0,\infty) \end{aligned} \tag{11.61}$$

Furthermore, we can obtain

$$\lim_{t\to\infty}|x(s,t)| \le \sqrt{\frac{L\varepsilon}{\lambda_1\lambda}}, \forall s \in [0, L] \tag{11.62}$$

$$\lim_{t\to\infty}|y(s,t)| \le \sqrt{\frac{L\varepsilon}{\lambda_1\lambda}}, \forall s \in [0, L] \tag{11.63}$$

$$\lim_{t\to\infty}|z(s,t)| \le \sqrt{\frac{L\varepsilon}{\lambda_1\lambda}}, \forall s \in [0, L] \tag{11.64}$$

Remark 11.2 When the parameters of the system are appropriately selected, we can adjust the control gains k_1, k_2 and k_3 to give out a better performance of the system. The increasing in the control gains k_1, k_2 and k_3 will result in a larger λ_3. Therefore the value of λ will increase, which will reduce the size of Ω_1 and Ω_2. Thus the bound of the system states $x(s,t)$, $y(s,t)$ and $z(s,t)$ can be arbitrarily small by selecting control parameters appropriately. However, increasing k_1, k_2 and k_3 will bring a high gain control problem. Therefore, in practical applications, the design parameters should be adjusted carefully for achieving suitable transient performance and control action.

Remark 11.3 We have $\zeta(t) \le \frac{1}{\lambda_1}V(t) \in \mathcal{L}_\infty$, we can state that $V_1(t)$ is bounded $\forall t \in [0,\infty)$. The knowledge of the exact boundary for $V_1(t)$ is not required, which is also suitable for other bounded signals mentioned below. Since $V_1(t)$ is bounded, $\dot{x}(s,t), x'(s,t), x''(s,t), \dot{y}(s,t)$, $y'(s,t)$, $y''(s,t)$, $\dot{z}(s,t)$ and $z'(s,t)$ are bounded $\forall(s,t) \in [0,L]\times[0,\infty)$. The kinetic energy $E_k(t)$ of the system is bounded, $\dot{x}'(s,t)$, $\dot{y}'(s,t)$ and $\dot{z}'(s,t)$ are bounded $\forall(s,t) \in [0,L]\times[0,\infty)$. We can conclude that $z''(s,t)$ is bounded $\forall(s,t) \in [0,L]\times[0,\infty)$. Considering Assumption 11.1 and the aforesaid statements, which means all signals on the right side of the governing equation are bounded. Then, we can conclude that $\ddot{x}(s,t)$, $\ddot{y}(s,t)$ and $\ddot{z}(s,t)$ defined in (6)–(8) are also bounded $\forall(s,t) \in [0,L]\times[0,\infty)$. From the proof of Theorem 1, it is shown the deflection $x(s,t)$, $y(s,t)$ and $z(s,t)$ are uniformly bounded $\forall(s,t) \in [0,L]\times[0,\infty)$. Since $\dot{x}(s,t)$, $\dot{y}(s,t)$, $\dot{z}(s,t)$, $x'(s,t)$, $y'(s,t)$, $z'(s,t)$, $\dot{x}'(s,t)$, $\dot{y}'(s,t)$ and $\dot{z}'(s,t)$ are all bounded $\forall(s,t) \in [0,L]\times[0,\infty)$, and we can accordingly conclude that the control inputs u_x, u_y and u_z are bounded $\forall t \in [0,\infty)$. Based on the above boundedness analysis, the proposed controller is effective and achievable in the actual system.

11.3.2 Adaptive Boundary Control to Compensate the Uncertainty of Tip Payload

The design of model-based controller relies on the systems mathematical model and the exact knowledge of all parameters of the system. Therefore, we ignore the uncertainty of tip payload M to design model-based controller in the former section. However, in the real situation, the uncertainty of M might exist. For example, when the manipulator working at an assembly line in a car factory picks up different parts of a car or other components, the mass of the payload is uncertain. Consequently, we further propose the adaptive control to improve the performance of the system via parameter estimation when there are some unknown parameters. In this section, the adaptive control is designed to compensate for the parameter uncertainty of M where the system parameter cannot be directly measured.

The adaptive control laws are design as follows

$$u_x = -\hat{M}(t)\beta L\dot{x}'_L - \mathrm{sgn}(\dot{x}_L + \beta L x'_L)\bar{d}_x - 2k_1\dot{x}_L \quad (11.65)$$

$$u_y = -\hat{M}(t)\beta L\dot{y}'_L - \mathrm{sgn}(\dot{y}_L + \beta L y'_L)\bar{d}_y - 2k_2\dot{y}_L \quad (11.66)$$

$$u_z = -\hat{M}(t)\beta L\dot{z}'_L - \mathrm{sgn}(\dot{z}_L + \beta L z'_L)\bar{d}_z - 2k_3\dot{z}_L \quad (11.67)$$

where $\hat{M}(t)$ is the estimate of M. We define $\tilde{M}(t) = M - \hat{M}(t)$ is the estimate error. The adaptive law is designed as

$$\dot{\hat{M}}(t) = \gamma\beta L[\dot{x}'_L(\dot{x}_L + \beta L x'_L) + \dot{y}'_L(\dot{y}_L + \beta L y'_L) + \dot{z}'_L(\dot{z}_L + \beta L z'_L)] - \gamma\xi\hat{M}(t) \quad (11.68)$$

where γ and ξ are two positive constants.

Remark 11.4 In the proposed controllers, $\dot{x}(L,t)$, $\dot{y}(L,t)$ and $\dot{z}(L,t)$ can be measured by the laser displacement sensors at top boundary of the beam, $x'(L,t)$, $y'(L,t)$ and $z'(L,t)$ can be measured through inclinometers, $\dot{x}(L,t), \dot{y}(L,t), \dot{z}(L,t)$, $\dot{x}'(L,t)$, $\dot{y}'(L,t)$ and $\dot{z}'(L,t)$ can be calculated with a backward difference algorithm.

Remark 11.5 The modification term ξ is introduced to improve the robustness of the closed-loop system. Without such a modification term, the estimation $\hat{M}(t)$ might drift to a very large value, which will result in a variation of a high-gain control scheme. In addition, bursting phenomena will be eliminated by using the modification term ξ.

Consider the following Lyapunov candidate function

$$V_A(t) = V(t) + \frac{1}{2}\gamma^{-1}\tilde{M}^2(t) \quad (11.69)$$

Lemma 11.3 *The Lyapunov candidate function can be upper and lower bounded as*

$$0 \leq \lambda_{1A}(\zeta(t) + V_2(t) + \tilde{M}(t)) \leq V_A(t) \leq \lambda_{2A}(\zeta(t) + V_2(t) + \tilde{M}(t)) \quad (11.70)$$

where $\lambda_{1A} = \min(\lambda_1, \frac{1}{2\gamma})$ *and* $\lambda_{2A} = \max(\lambda_2, \frac{1}{2\gamma})$ *are two positive constants.*

Proof We can get

$$0 \leq \lambda_{1A}(\zeta(t) + V_2(t) + \tilde{M}(t)) \leq V_A(t) \leq \lambda_{2A}(\zeta(t) + V_2(t) + \tilde{M}(t)) \quad (11.71)$$

where $\lambda_{1A} = \min(\lambda_1, \frac{1}{2\gamma})$ and $\lambda_{2A} = \max(\lambda_2, \frac{1}{2\gamma})$ are two positive constants.

Lemma 11.4 *The time derivative of the Lyapunov candidate function can be upper bounded with*

$$\dot{V}_A(t) \leq -\lambda_A V_A(t) + \varepsilon_A, \quad (11.72)$$

where λ_A *and* ε_A *are two positive constants.*

Proof With the adaptive control laws, we can rewrite Δ as

$$\begin{aligned}\Delta_A \leq &- k_1(\dot{x}_L + \beta L x_L')^2 - k_2(\dot{y}_L + \beta L y_L')^2 - k_3(\dot{z}_L + \beta L z_L')^2 \\ &- \Big(k_1 - \frac{1}{2}\beta\rho L\Big)(\dot{x}_L)^2 - \Big(k_2 - \frac{1}{2}\beta\rho L\Big)(\dot{y}_L)^2 - \Big(k_3 - \frac{1}{2}\beta\rho L\Big)(\dot{z}_L)^2 \\ &- \Big(\frac{1}{4}\beta T L - k_1\beta^2 L^2\Big)(x_L')^2 - \Big(\frac{1}{4}\beta T L - k_2\beta^2 L^2\Big)(y_L')^2 - \Big(\beta T L - k_3\beta^2 L^2\Big)(z_L')^2 \\ &- \frac{1}{2}\beta E A L\{(z_L')^2 + z_L'[(x_L')^2 + (y_L')^2] + \frac{1}{4}[(x_L')^2 + (y_L')^2]^2\} \\ &+ \tilde{M}(t)[\beta L\dot{x}_L'(\dot{x}_L + \beta L x_L') + \beta L\dot{y}_L'(\dot{y}_L + \beta L y_L') + \beta L\dot{z}_L'(\dot{z}_L + \beta L z_L')] \end{aligned} \quad (11.73)$$

Comparing with Δ, the last term, $\tilde{M}(t)[\beta L\dot{x}_L'(\dot{x}_L + \beta L x_L') + \beta L\dot{y}_L'(\dot{y}_L + \beta L y_L') + \beta L\dot{z}_L'(\dot{z}_L + \beta L z_L')]$, of Δ_A is caused by $\hat{M}(t)$. Combining the adaptive law, the time derivative of $\frac{1}{2}\gamma^{-1}\tilde{M}^2(t)$ can be written as

$$\begin{aligned}\gamma^{-1}\tilde{M}(t)\dot{\tilde{M}}(t) = &- \tilde{M}(t)[\beta L\dot{x}_L'(\dot{x}_L + \beta L x_L') + \beta L\dot{y}_L'(\dot{y}_L + \beta L y_L') \\ &+ \beta L\dot{z}_L'(\dot{z}_L + \beta L z_L')] + \xi\tilde{M}(t)\hat{M}(t) \end{aligned} \quad (11.74)$$

which can eliminate the last term in Δ_A to $\xi\tilde{M}(t)\hat{M}(t)$.

Therefore, we obtain the time derivative of $V_A(t)$ as

$$\begin{aligned}\dot{V}_A(t) =&\dot{V}(t) + \gamma^{-1}\tilde{M}(t)\dot{\tilde{M}}(t)\\ \leq& -\lambda_3[\zeta(t) + V_2(t)] + \varepsilon + \xi\tilde{M}(t)\hat{M}(t)\\ \leq& -\lambda_3[\zeta(t) + V_2(t)] + \varepsilon - \frac{\xi}{2}\tilde{M}^2(t) + \frac{\xi}{2}M^2\\ \leq& -\lambda_{3A}[\zeta(t) + V_2(t) + \tilde{M}^2(t)] + \varepsilon_A\\ \leq& -\lambda_A V_A(t) + \varepsilon_A\end{aligned} \tag{11.75}$$

where $\lambda_{3A} = \min(\lambda_3, \frac{\xi}{2})$, $\lambda_A = \lambda_{3A}/\lambda_{2A}$, $\varepsilon_A = \varepsilon + \frac{\xi}{2}M^2 \in \mathcal{L}_\infty$.

With Lemma 11.2, we are ready to present the following stability theorem for the Euler–Bernoulli beam system.

Theorem 11.2 *For the system and boundary conditions, under Assumption 11.1, and the control laws, given that the initial conditions are bounded, we can conclude that*

(i) uniform boundedness (UB): the state of the system $x(s,t)$, $y(s,t)$ and $z(s,t)$ will remain in the compact set Ω_3 defined by $\Omega_3 := \{x(s,t), y(s,t), z(s,t) \in R|\ |x(s,t)|, |y(s,t)|, |z(s,t)| \leq D_3, \forall(s,t) \in [0,L] \times [0,\infty)\}$, where the constant $D_3 = \sqrt{\frac{L}{\lambda_{1A}}\left(V(0) + \frac{\varepsilon_A}{\lambda}\right)}$.

(ii) uniform ultimate boundedness (UUB):the state of the system $x(s,t)$, $y(s,t)$ and $z(s,t)$ will eventually converge to the compact Ω_4 defined by $\Omega_4 := \{x(s,t), y(s,t), z(s,t) \in R|\quad \lim_{t\to\infty}|x(s,t)|, \lim_{t\to\infty}, \quad |y(s,t)|, \lim_{t\to\infty}|z(s,t)| \leq D_2, \forall t \in [0,\infty)\}$, where the constant $D_4 = \sqrt{\frac{L\varepsilon_A}{\lambda_{1A}\lambda_A}}$.

Proof

$$\frac{d}{dt}(V_A e^{\lambda_A t}) \leq \varepsilon_A e^{\lambda_A t} \tag{11.76}$$

Integrating of the above inequality, we obtain

$$V_A(t) \leq \left(V_A(0) - \frac{\varepsilon_A}{\lambda_A}\right)e^{-\lambda_A t} + \frac{\varepsilon_A}{\lambda_A} \leq V(0)e^{-\lambda_A t} + \frac{\varepsilon_A}{\lambda_A} \in \mathcal{L}_\infty \tag{11.77}$$

which implies $\dot{V}(t)$ is bounded. We obtain

$$\frac{1}{L}[x(s,t)]^2 \leq \int_0^L [x'(s,t)]^2 ds \leq \zeta(t) \leq \frac{1}{\lambda_{1A}}V_A(t) \in \mathcal{L}_\infty \tag{11.78}$$

$$\frac{1}{L}[y(s,t)]^2 \leq \int_0^L [y'(s,t)]^2 ds \leq \zeta(t) \leq \frac{1}{\lambda_{1A}}V_A(t) \in \mathcal{L}_\infty \tag{11.79}$$

$$\frac{1}{L}[z(s,t)]^2 \leq \int_0^L [z'(s,t)]^2 ds \leq \zeta(t) \leq \frac{1}{\lambda_{1A}}V_A(t) \in \mathcal{L}_\infty \tag{11.80}$$

Rearranging the terms of the above three inequalities, we can obtain

$$|x(s,t)|, |y(s,t)|, |z(s,t)| \leq \sqrt{\frac{L}{\lambda_{1A}}\left(V(0)e^{-\lambda_A t} + \frac{\varepsilon_A}{\lambda_A}\right)} \tag{11.81}$$

Furthermore, we can obtain

$$\lim_{t\to\infty} |x(s,t)| \leq \sqrt{\frac{L\varepsilon_A}{\lambda_{1A}\lambda_A}} \tag{11.82}$$

$$\lim_{t\to\infty} |y(s,t)| \leq \sqrt{\frac{L\varepsilon_A}{\lambda_{1A}\lambda_A}} \tag{11.83}$$

$$\lim_{t\to\infty} |z(s,t)| \leq \sqrt{\frac{L\varepsilon_A}{\lambda_{1A}\lambda_A}} \tag{11.84}$$

Remark 11.6 We can conclude that system state $x(s,t)$, $y(s,t)$ and $z(s,t)$ with the proposed adaptive control can be made arbitrarily small by choosing appropriate control gains k_1, k_2 and k_3.

Remark 11.7 We obtain that the estimate error $\tilde{M}$ is bounded $\forall t \in [0, \infty)$. Using analysis similar to Remark 4, we can conclude $\dot{x}(s,t)$, $\ddot{x}(s,t)$, $x'(s,t)$, $x''(s,t)$, $\dot{x}'(s,t)$, $\dot{y}(s,t)$, $\ddot{y}(s,t)$, $y'(s,t)$, $y''(s,t)$, $\dot{y}'(s,t)$, $\dot{z}(s,t)$, $\ddot{z}(s,t)$, $z'(s,t)$ and $z''(s,t)$ are all bounded $\forall(s,t) \in [0, L] \times [0, \infty)$. Therefore, the deflections $x(s,t)$, $y(s,t)$ and $z(s,t)$ are uniformly bounded $\forall(s,t) \in [0, L] \times [0, \infty)$. And the adaptive boundary control are bounded $\forall t \in [0, \infty)$.

11.4 Numerical Simulations

In this section, in order to demonstrate the effectiveness and practicability of the proposed model-based boundary control laws and adaptive boundary control laws, we choose the finite differential method to carry out the numerical simulation. The beam is influenced by the spatiotemporally environmental disturbances which are defined as

$$f_x(s,t) = f_y(s,t) = \frac{s}{5L}[1 + 2\sin(0.2st) + 3\sin(0.3st) + 5\sin(0.5st)] \tag{11.85}$$

$$f_z(s,t) = \frac{s}{5L}[2 + 2\sin(0.2st) + 3\sin(0.3st) + 5\sin(0.5st)] \tag{11.86}$$

$$d_x(t) = d_y(t) = d_z(t) = \frac{1}{5}[1 + 4\sin(0.4t)] \tag{11.87}$$

The initial conditions of the beam system are described as

$$x(s,0) = y(s,0) = z(s,0) = \frac{s}{2L} \tag{11.88}$$

$$\dot{x}(s,0) = \dot{y}(s,0) = \dot{z}(s,0) = 0 \tag{11.89}$$

Parameters of the system are given as follows (Table 11.1).

When there is no control applied to the system, the beam is freely vibrated and the transverse deflections of the system in X and Y directions can be close to 0.6 m shown in Figs. 11.2 and 11.3, while the longitudinal deflection in Z direction can reach and even exceed 0.6 m depicted in Fig. 11.4. In Figs. 11.5 and 11.6, we can see that when the model-based controller is applied, the vibration amplitudes of the system in transverse directions can be greatly regulated in about 5 s, similarly, the significant vibration in the longitudinal direction observed in Fig. 11.7 can be reduced to the small neighborhood of zero within 6 s by choosing the appropriate control parameters as $\beta = 0.5$, $k_1 = k_2 = 15$ and $k_3 = 5$. From Figs. 11.12, 11.13 and 11.14, we can see that when the adaptive control is applied, the transverse and longitudinal displacement amplitudes of the system in X, Y and Z directions can be greatly regulated in about 4 and 5 s, respectively. Comparing with the model-based control, the parameters of adaptive control are designed as $\beta = 0.5$, $k_1 = k_2 = 10$, $k_3 = 5$, $\gamma = 2$ and $\xi = 1$. Therefore, we can conclude that the proposed boundary controls are able to stabilize the beam system at the small neighborhood of zero by appropriately choosing different parameters in three directions. The control inputs for the model-based control and adaptive control are shown in Figs. 11.11 and 11.18, respectively. The endpoint deflections under the model-based control and adaptive control are shown in Figs. 11.8, 11.9, 11.10, 11.15, 11.16 and 11.17.

Table 11.1 Parameters of the system

Parameter	Discription	Value
L	Length of beam	1 m
ρ	Uniform mass per unit length of the beam	0.1 kg/m
M	Mass of the tip payload	1 kg
EI	Bending stiffness of the beam	8 Nm^2
EA	Axial stiffness of the beam	14 Nm^2
T	Tension of the beam	30 N

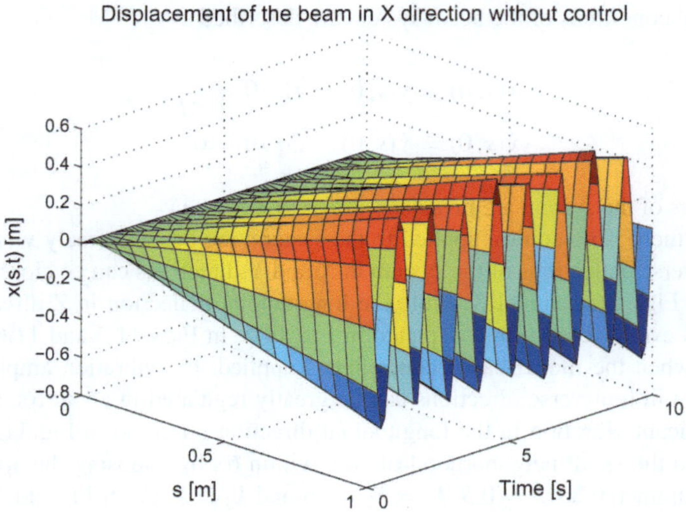

Fig. 11.2 Displacement of the beam in the X direction without control

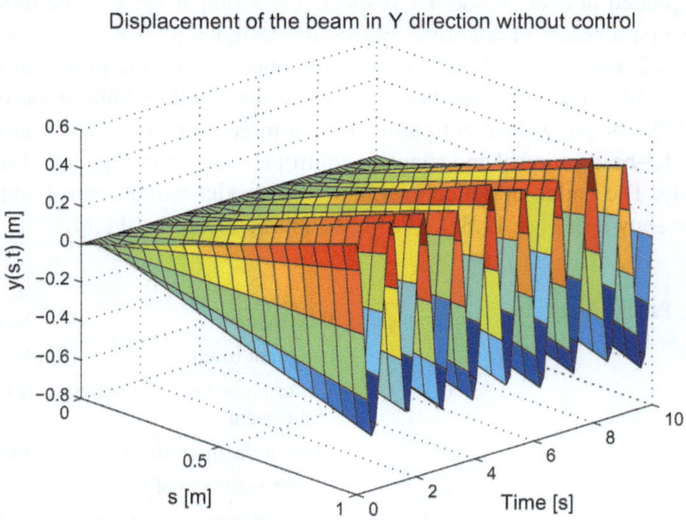

Fig. 11.3 Displacement of the beam in the Y direction without control

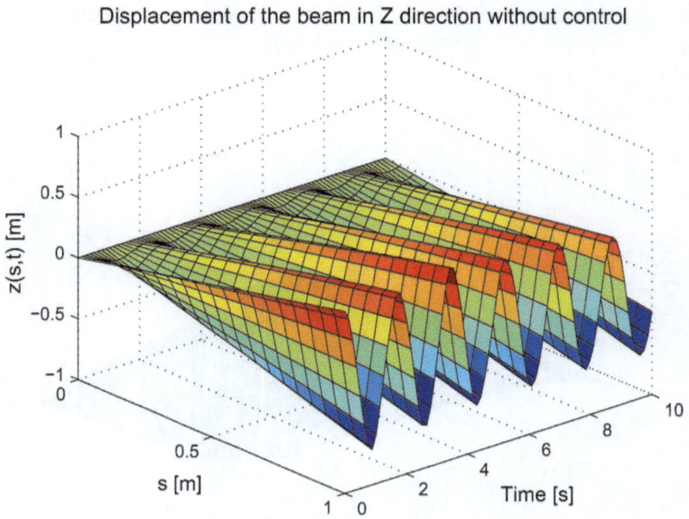

Fig. 11.4 Displacement of the beam in the Z direction without control

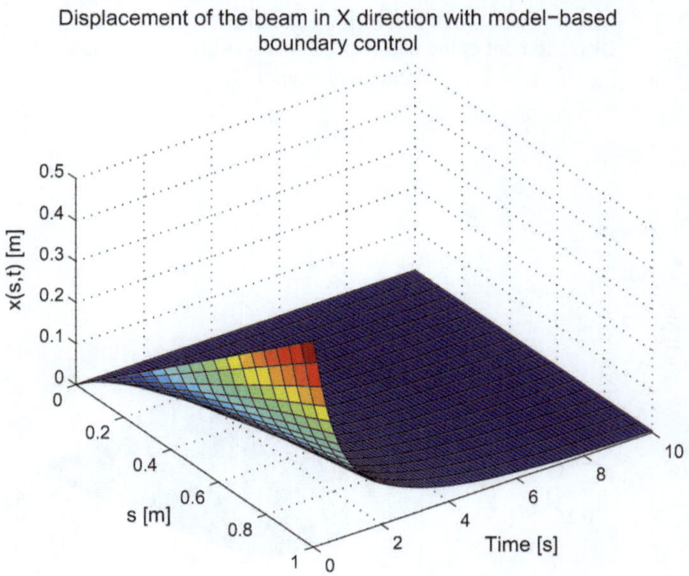

Fig. 11.5 Displacement of the beam at the X direction with boundary control

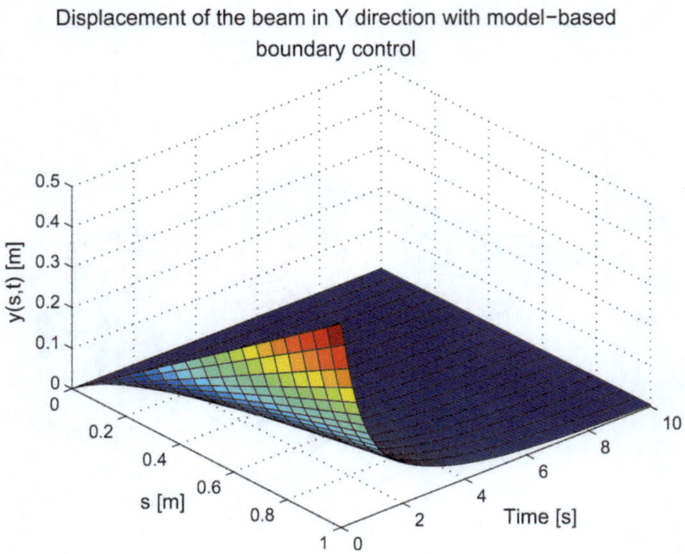

Fig. 11.6 Displacement of the beam at the Y direction with boundary control

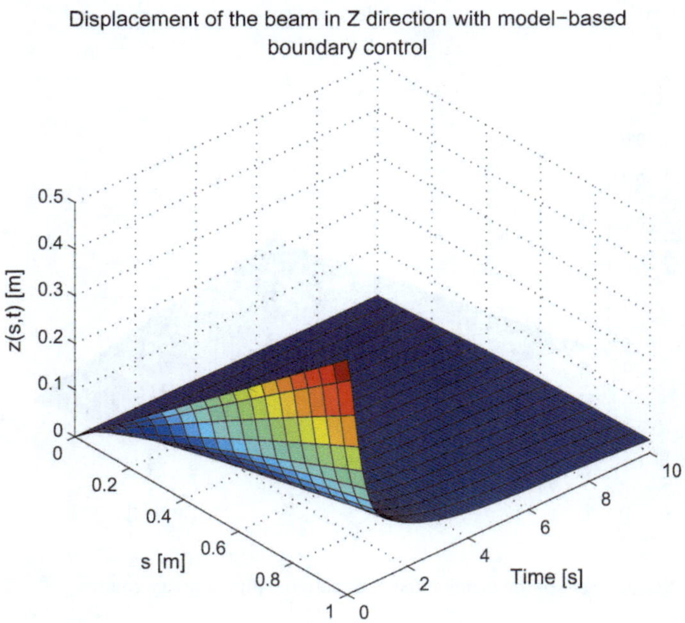

Fig. 11.7 Displacement of the beam at the Z direction with boundary control

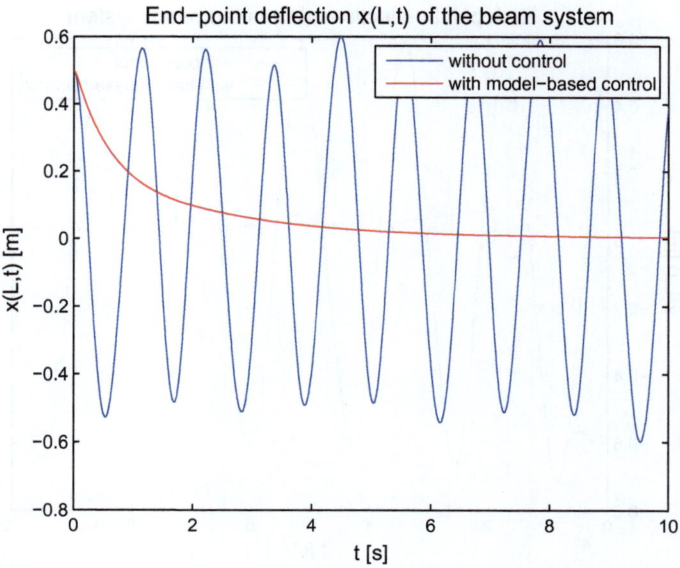

Fig. 11.8 Displacement of the beam's endpoint at the X direction with boundary control

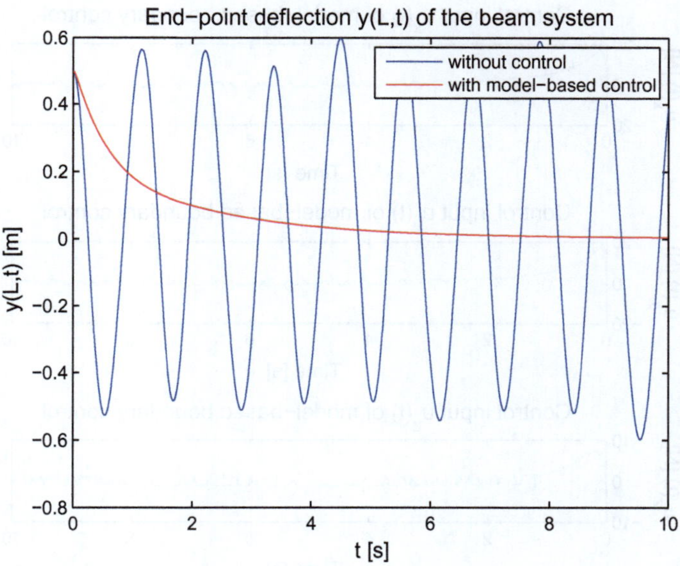

Fig. 11.9 Displacement of the beam's endpoint at the Y direction with boundary control

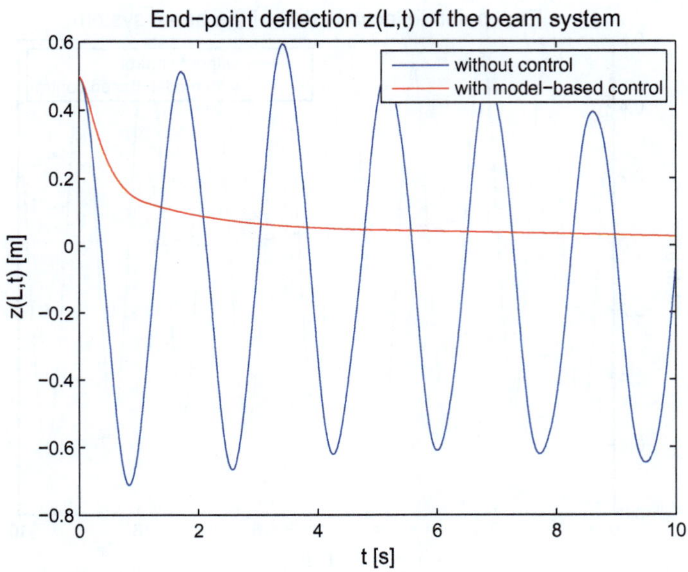

Fig. 11.10 Displacement of the beam's endpoint at the Z direction with boundary control

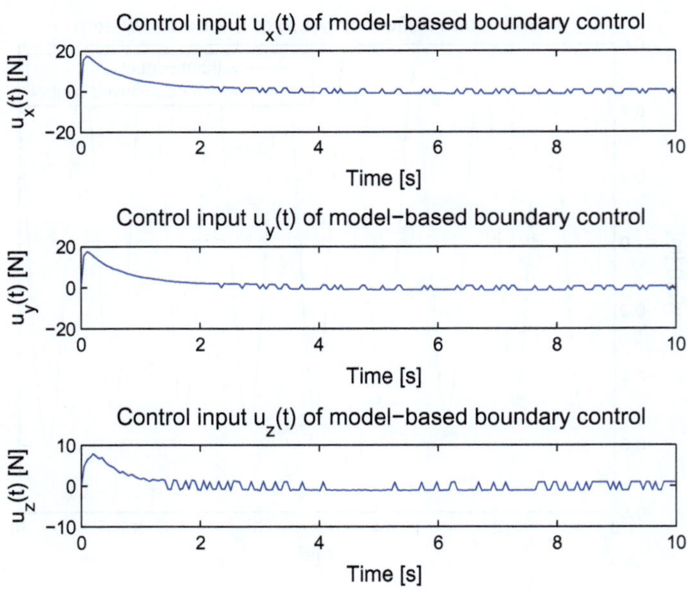

Fig. 11.11 Model-based control input

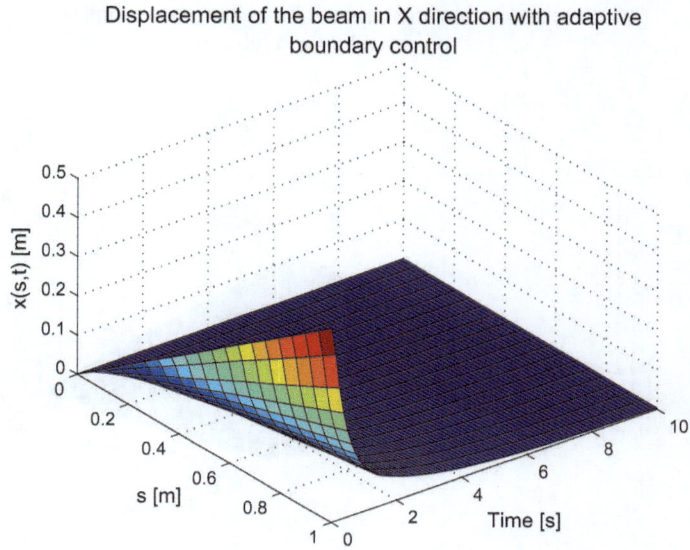

Fig. 11.12 Displacement of the beam at the X direction with adaptive boundary control

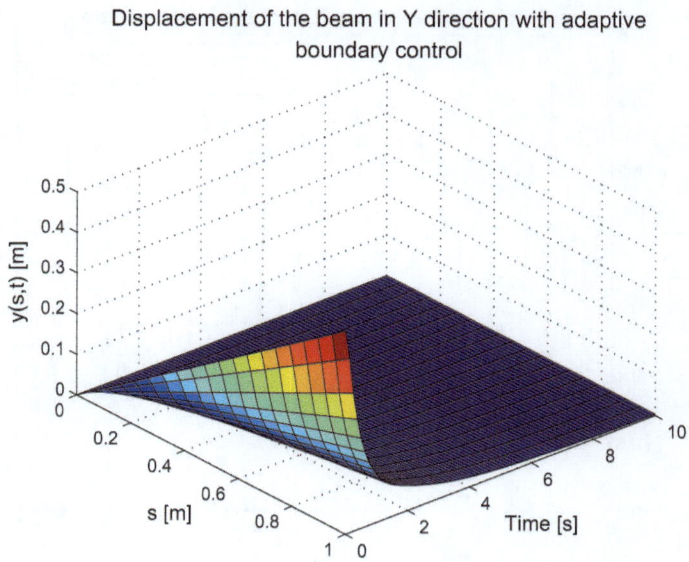

Fig. 11.13 Displacement of the beam at the Y direction with adaptive boundary control

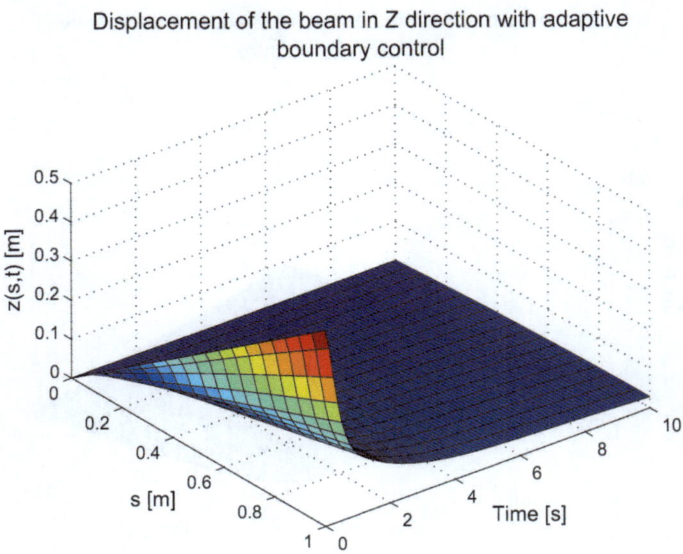

Fig. 11.14 Displacement of the beam at the Z direction with adaptive boundary control

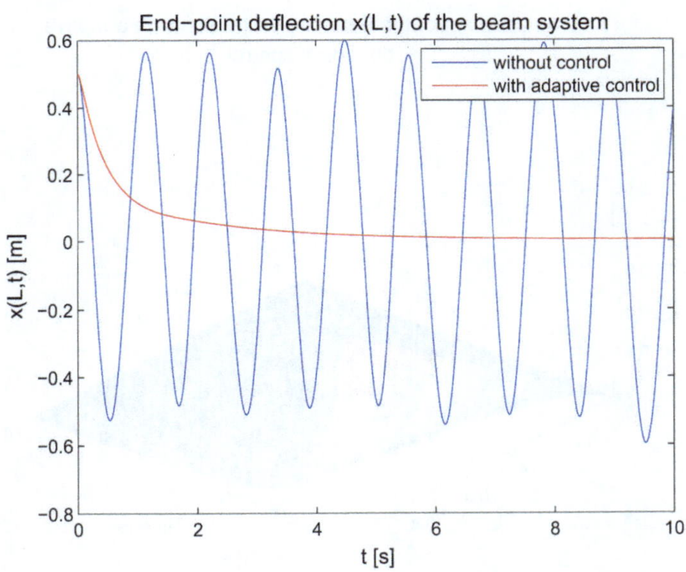

Fig. 11.15 Displacement of the beam's endpoint at the X direction with adaptive boundary control

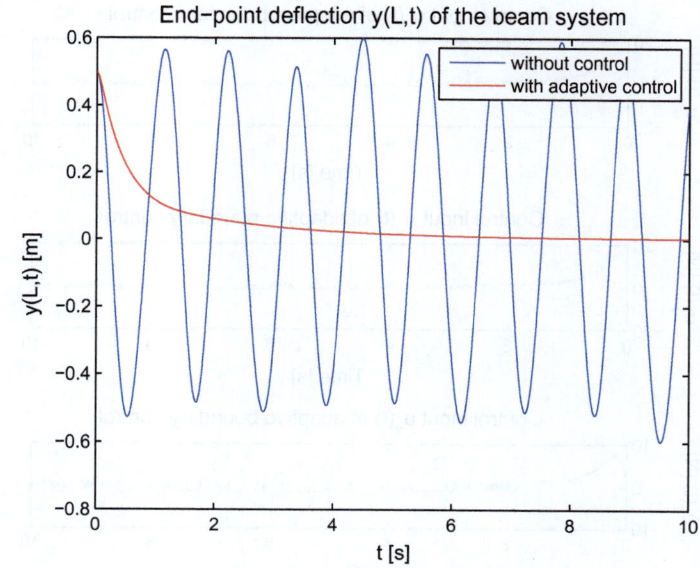

Fig. 11.16 Displacement of the beam's endpoint at the Y direction with adaptive boundary control

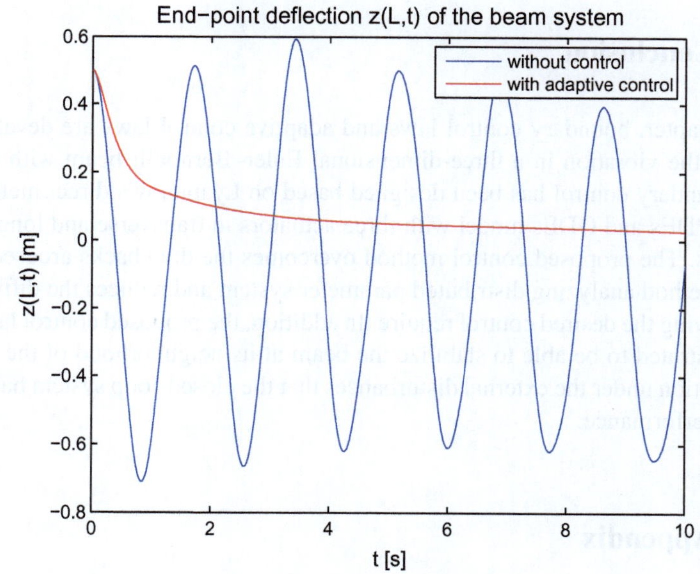

Fig. 11.17 Displacement of the beam's endpoint at the Z direction with adaptive boundary control

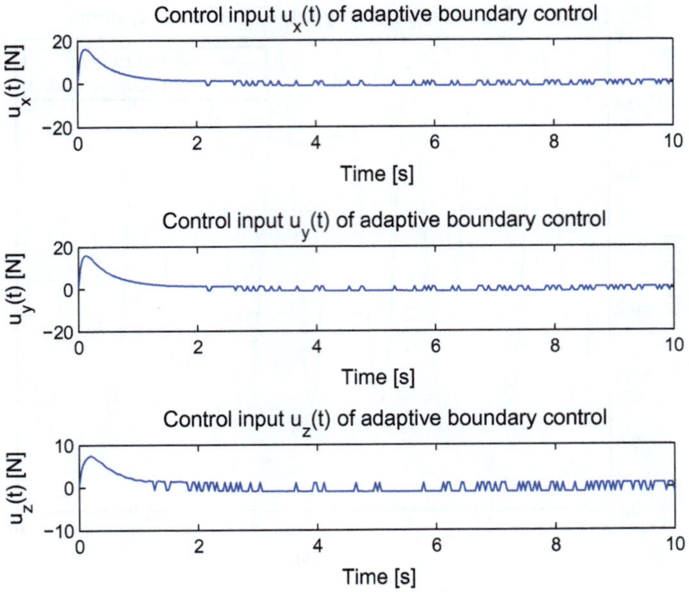

Fig. 11.18 Adaptive control input

11.5 Conclusion

In this chapter, boundary control laws and adaptive control laws are developed to suppress the vibration in a three-dimensional Euler–Bernoulli beam with tip payload. Boundary control has been designed based on Lyapunov's direct method and original PDEs and ODEs model with three actuators in transverse and longitudinal directions. The proposed control method overcomes the drawbacks aroused by the typical method analyzing distributed parameter system and reduces the difficulty of the achieving the desired control require. In addition, the proposed control laws have been illustrated to be able to stabilize the beam at its neighborhood of the equilibrium position under the external disturbances that the closed-loop system has a good control performance.

11.6 Appendix

```
clc;
close all;
clear all;
%%
```

```
%*********************************************************
%                     3D beam Model-based
%*********************************************************
nx=20;
nt=80*10^3;

L=1;
tmax=10;
Ttr=80;

h=L/(nx-1);
k=tmax/(nt-1);

% constants
M=1;
EI=8;
EA=14;
T=30;
rho=0.1;
beta=0.5;

k1=15;
k2=15;
k3=5;

fx=zeros(nx,nt);
fy=zeros(nx,nt);
fz=zeros(nx,nt);

dx=zeros(nt,1);
dy=zeros(nt,1);
dz=zeros(nt,1);

for j=1:nt
    for i=1:nx
        fx(i,j)=( 1+2*sin(0.2*(i-1)*h*j*k)+3*sin(0.3*(i-1)*h*j*k)
                +5*sin(0.5*(i-1)*h*j*k) )*(i-1)*h/L/5;
        fy(i,j)=( 1+2*sin(0.2*(i-1)*h*j*k)+3*sin(0.3*(i-1)*h*j*k)
                +5*sin(0.5*(i-1)*h*j*k) )*(i-1)*h/L/5;
        fz(i,j)=( 2+2*sin(0.2*(i-1)*h*j*k)+3*sin(0.3*(i-1)*h*j*k)
                +5*sin(0.5*(i-1)*h*j*k) )*(i-1)*h/L/5;
    end

    dx(j)=(1+4*sin(0.4*(j-1)*k) )/5;
    dy(j)=(1+4*sin(0.4*(j-1)*k) )/5;
```

```
    dz(j)=(1+4*sin(0.4*(j-1)*k) )/5;
end

% create matrixes to save state data
x=zeros(nx,nt);
y=zeros(nx,nt);
z=zeros(nx,nt);

% create matrixes to save control inputs
ux=zeros(nt,1);
uy=zeros(nt,1);
uz=zeros(nt,1);

x_3D_free=zeros(Ttr,nx);
y_3D_free=zeros(Ttr,nx);
z_3D_free=zeros(Ttr,nx);

xl_2D_free=zeros(nt,1);
yl_2D_free=zeros(nt,1);
zl_2D_free=zeros(nt,1);

% initial condition

for i=1:nx
    x(i,1)=0.5*(i-1)*h/L;
    y(i,1)=0.5*(i-1)*h/L;
    z(i,1)=0.5*(i-1)*h/L;
end

x(:,2)=x(:,1);
y(:,2)=y(:,1);
z(:,2)=z(:,1);

% main cycle
for j=2:nt-1

x(1,j)=0;
y(1,j)=0;
z(1,j)=0;

    for i=3:nx-2
        xs=( x(i,j)-x(i-1,j) )/h;
        ys=( y(i,j)-y(i-1,j) )/h;
        zs=( z(i,j)-z(i-1,j) )/h;
        xss=( x(i+1,j)-2*x(i,j)+x(i-1,j) )/h^2;
```

```
        yss=( y(i+1,j)-2*y(i,j)+y(i-1,j) )/h^2;
        zss=( z(i+1,j)-2*z(i,j)+z(i-1,j) )/h^2;
        xssss=( x(i+2,j)-4*x(i+1,j)+6*x(i,j)-4*x(i-1,j)
                +x(i-2,j) )/h^4;
        yssss=( y(i+2,j)-4*y(i+1,j)+6*y(i,j)-4*y(i-1,j)
                +y(i-2,j) )/h^4;
        S1=T*xss-EI*xssss+EA*(zss*xs+xss*zs)+1.5*EA*xs^2*xss
            +0.5*EA*( xss*ys^2+2*xs*ys*yss )+fx(i,j);
        S2=T*yss-EI*yssss+EA*(zss*ys+yss*zs)+1.5*EA*ys^2*yss
            +0.5*EA*( yss*xs^2+2*ys*xs*xss )+fy(i,j);
        S3=EA*zss+EA*xs*xss+EA*ys*yss+fz(i,j);
        x(i,j+1)=2*x(i,j)-x(i,j-1)+k^2*S1/rho;
        y(i,j+1)=2*y(i,j)-y(i,j-1)+k^2*S2/rho;
        z(i,j+1)=2*z(i,j)-z(i,j-1)+k^2*S3/rho;
    end

    x(2,j+1)=(x(1,j+1)+x(3,j+1))/2;
    y(2,j+1)=(y(1,j+1)+y(3,j+1))/2;
    z(2,j+1)=(z(1,j+1)+z(3,j+1))/2;

        xsl=(x(nx,j)-x(nx-1,j))/h;
        ysl=(y(nx,j)-y(nx-1,j))/h;
        zsl=(z(nx,j)-z(nx-1,j))/h;
        xsssl=(-x(nx,j)+2*x(nx-1,j)-x(nx-2,j))/h^3;
        ysssl=(-y(nx,j)+2*y(nx-1,j)-y(nx-2,j))/h^3;

        x(nx,j+1)=2*x(nx,j)-x(nx,j-1)-( T*xsl+0.5*EA*xsl^3
                  +EA*xsl*zsl+0.5*EA*xsl*ysl^2-EI*xsssl
                  -dx(j)-ux(j) )*k^2/M;
        y(nx,j+1)=2*y(nx,j)-y(nx,j-1)-( T*ysl+0.5*EA*ysl^3
                  +EA*ysl*zsl+0.5*EA*ysl*xsl^2-EI*ysssl
                  -dy(j)-uy(j) )*k^2/M;
        z(nx,j+1)=2*z(nx,j)-z(nx,j-1)-( EA*zsl+0.5*EA*xsl^2
                  +0.5*EA*ysl^2-dz(j)-uz(j) )*k^2/M;

    x(nx-1,j+1)=( x(nx,j+1)+x(nx-2,j+1) )/2;
    y(nx-1,j+1)=( y(nx,j+1)+y(nx-2,j+1) )/2;
    z(nx-1,j+1)=( z(nx,j+1)+z(nx-2,j+1) )/2;


xl_2D_free(j+1)=x(nx,j+1);
yl_2D_free(j+1)=y(nx,j+1);
zl_2D_free(j+1)=z(nx,j+1);
```

```
    if mod(j,nt/Ttr)==0
        x_3D_free(1+j*Ttr/nt,:)=x(:,j)';
        y_3D_free(1+j*Ttr/nt,:)=y(:,j)';
        z_3D_free(1+j*Ttr/nt,:)=z(:,j)';
    end
end

x_3D_free(1,:)=x(:,1)';
y_3D_free(1,:)=y(:,1)';
z_3D_free(1,:)=z(:,1)';

xl_2D_free(1)=x(nx,1);
xl_2D_free(2)=xl_2D_free(1);
yl_2D_free(1)=y(nx,1);
yl_2D_free(2)=yl_2D_free(1);
zl_2D_free(1)=z(nx,1);
zl_2D_free(2)=zl_2D_free(1);


%%
% ****************************************************************
%     with control, i.e., model-based boundary control
% ****************************************************************
Ttr1=200;
dx_bar=max(max(dx));
dy_bar=max(max(dy));
dz_bar=max(max(dz));

% create matrixes to save state data
x=zeros(nx,nt);
y=zeros(nx,nt);
z=zeros(nx,nt);

% create matrixes to save control inputs
ux=zeros(nt,1);
uy=zeros(nt,1);
uz=zeros(nt,1);

x_3D_control=zeros(Ttr,nx);
y_3D_control=zeros(Ttr,nx);
z_3D_control=zeros(Ttr,nx);

xl_2D_control=zeros(nt,1);
yl_2D_control=zeros(nt,1);
zl_2D_control=zeros(nt,1);
```

```
ux_2D=zeros(Ttr1,1);
uy_2D=ux_2D;
uz_2D=ux_2D;


for i=1:nx
    x(i,1)=0.5*(i-1)*h/L;
    y(i,1)=0.5*(i-1)*h/L;
    z(i,1)=0.5*(i-1)*h/L;
end

x(:,2)=x(:,1);
y(:,2)=y(:,1);
z(:,2)=z(:,1);

% main cycle
for j=2:nt-1

x(1,j)=0;
y(1,j)=0;
z(1,j)=0;

    for i=3:nx-2

        xs=( x(i,j)-x(i-1,j) )/h;
        ys=( y(i,j)-y(i-1,j) )/h;
        zs=( z(i,j)-z(i-1,j) )/h;
        xss=( x(i+1,j)-2*x(i,j)+x(i-1,j) )/h^2;
        yss=( y(i+1,j)-2*y(i,j)+y(i-1,j) )/h^2;
        zss=( z(i+1,j)-2*z(i,j)+z(i-1,j) )/h^2;
        xssss=( x(i+2,j)-4*x(i+1,j)+6*x(i,j)-4*x(i-1,j)
            +x(i-2,j) )/h^4;
        yssss=( y(i+2,j)-4*y(i+1,j)+6*y(i,j)-4*y(i-1,j)
            +y(i-2,j) )/h^4;
        S1=T*xss-EI*xssss+EA*(zss*xs+xss*zs)+1.5*EA*xs^2*xss
           +0.5*EA*( xss*ys^2+2*xs*ys*yss )+fx(i,j);
        S2=T*yss-EI*yssss+EA*(zss*ys+yss*zs)+1.5*EA*ys^2*yss
           +0.5*EA*( yss*xs^2+2*ys*xs*xss )+fy(i,j);
        S3=EA*zss+EA*xs*xss+EA*ys*yss+fz(i,j);
        x(i,j+1)=2*x(i,j)-x(i,j-1)+k^2*S1/rho;
        y(i,j+1)=2*y(i,j)-y(i,j-1)+k^2*S2/rho;
          z(i,j+1)=2*z(i,j)-z(i,j-1)+k^2*S3/rho;
    end
```

```
x(2,j+1)=(x(1,j+1)+x(3,j+1))/2;
y(2,j+1)=(y(1,j+1)+y(3,j+1))/2;
z(2,j+1)=(z(1,j+1)+z(3,j+1))/2;

xsl=(x(nx,j)-x(nx-1,j))/h;
    ysl=(y(nx,j)-y(nx-1,j))/h;
    zsl=(z(nx,j)-z(nx-1,j))/h;
    xsssl=(-x(nx,j)+2*x(nx-1,j)-x(nx-2,j))/h^3;
    ysssl=(-y(nx,j)+2*y(nx-1,j)-y(nx-2,j))/h^3;

 x(nx,j+1)=2*x(nx,j)-x(nx,j-1)-( T*xsl+0.5*EA*xsl^3+EA*xsl*zsl
            +0.5*EA*xsl*ysl^2-EI*xsssl-dx(j)-ux(j) )*k^2/M;
 y(nx,j+1)=2*y(nx,j)-y(nx,j-1)-( T*ysl+0.5*EA*ysl^3+EA*ysl*zsl
            +0.5*EA*ysl*xsl^2-EI*ysssl-dy(j)-uy(j) )*k^2/M;
 z(nx,j+1)=2*z(nx,j)-z(nx,j-1)-( EA*zsl+0.5*EA*xsl^2
            +0.5*EA*ysl^2-dz(j)-uz(j) )*k^2/M;

        x(nx-1,j+1)=( x(nx,j+1)+x(nx-2,j+1) )/2;
y(nx-1,j+1)=( y(nx,j+1)+y(nx-2,j+1) )/2;
z(nx-1,j+1)=( z(nx,j+1)+z(nx-2,j+1) )/2;

% compute controller
    dxsl=( x(nx,j+1)-x(nx-1,j+1)-x(nx,j)+x(nx-1,j) )/(k*h);
    dysl=( y(nx,j+1)-y(nx-1,j+1)-y(nx,j)+y(nx-1,j) )/(k*h);
    dzsl=( z(nx,j+1)-z(nx-1,j+1)-z(nx,j)+z(nx-1,j) )/(k*h);
    dxl=( x(nx,j+1)-x(nx,j) )/k;
    dyl=( y(nx,j+1)-y(nx,j) )/k;
    dzl=( z(nx,j+1)-z(nx,j) )/k;
    xsl=(x(nx,j+1)-x(nx-1,j+1))/h;
    ysl=(y(nx,j+1)-y(nx-1,j+1))/h;
    zsl=(z(nx,j+1)-z(nx-1,j+1))/h;

 ux(j+1)=-M*beta*L*dxsl-sgn1( dxl+beta*L*xsl )*dx_bar-2*k1*dxl;
 uy(j+1)=-M*beta*L*dysl-sgn1( dyl+beta*L*ysl )*dy_bar-2*k2*dyl;
 uz(j+1)=-M*beta*L*dzsl-sgn1( dzl+beta*L*zsl )*dy_bar-2*k3*dzl;

xl_2D_control(j+1)=x(nx,j+1);
yl_2D_control(j+1)=y(nx,j+1);
zl_2D_control(j+1)=z(nx,j+1);


  if mod(j,nt/Ttr)==0
    x_3D_control(1+j*Ttr/nt,:)=x(:,j)';
    y_3D_control(1+j*Ttr/nt,:)=y(:,j)';
    z_3D_control(1+j*Ttr/nt,:)=z(:,j)';
```

```
    end

    if mod(j,nt/Ttr1)==0
        ux_2D(1+j*Ttr1/nt)=ux(j+1);
        uy_2D(1+j*Ttr1/nt)=uy(j+1);
        uz_2D(1+j*Ttr1/nt)=uz(j+1);
    end

end

x_3D_control(1,:)=x(:,1)';
y_3D_control(1,:)=y(:,1)';
z_3D_control(1,:)=z(:,1)';

xl_2D_control(1)=x(nx,1);
xl_2D_control(2)=xl_2D_control(1);
yl_2D_control(1)=y(nx,1);
yl_2D_control(2)=yl_2D_control(1);
zl_2D_control(1)=z(nx,1);
zl_2D_control(2)=zl_2D_control(1);


% save data for discussion about internal fluid
x0=linspace(0,L,nx);
t_tr=linspace(0,tmax,Ttr);

figure(1);
surf(x0,t_tr,x_3D_free);
view(45,25);
title({'Displacement of the beam in X direction without control'});
ylabel('Time s]','Fontsize',8);
xlabel('s [m]','Fontsize',8);
zlabel('x(s,t) [m]','Fontsize',8);

figure(2);
surf(x0,t_tr,y_3D_free);
view(45,25);
title({'Displacement of the beam in Y direction without control'});
ylabel('Time [s]','Fontsize',8);
xlabel('s [m]','Fontsize',8);
zlabel('y(s,t) [m]','Fontsize',8);

figure(3);
surf(x0,t_tr,z_3D_free);
view(45,30);
```

```
title({'Displacement of the beam in Z direction without control'});
ylabel('Time [s]','Fontsize',8);
xlabel('s [m]','Fontsize',8);
zlabel('z(s,t) [m]','Fontsize',8);

figure(4);
surf(x0,t_tr,x_3D_control);
view(50,40);
title({'Displacement of the beam in X direction with model-based 
    boundary control'});
ylabel('Time [s]','Fontsize',8);
xlabel('s [m]','Fontsize',8);
zlabel('x(s,t) [m]','Fontsize',8);

figure(5);
surf(x0,t_tr,y_3D_control);
view(50,40);
title({'Displacement of the beam in Y direction with model-based 
    boundary control'});
ylabel('Time [s]','Fontsize',8);
xlabel('s [m]','Fontsize',8);
zlabel('y(s,t) [m]','Fontsize',8);

figure(6);
surf(x0,t_tr,z_3D_control);
view(50,40);
title({'Displacement of the beam in Z direction with model-based 
    boundary control'});
ylabel('Time [s]','Fontsize',8);
xlabel('s [m]','Fontsize',8);
zlabel('z(s,t) [m]','Fontsize',8);

figure(7);
plot(linspace(0,tmax,nt),xl_2D_free,'b');
hold on
plot(linspace(0,tmax,nt),xl_2D_control,'r');
xlabel('t [s]');
ylabel('x(L,t) [m]');
title('End-point deflection x(L,t) of the beam system');
legend('without control','with model-based control');

figure(8);
plot(linspace(0,tmax,nt),yl_2D_free,'b');
hold on
plot(linspace(0,tmax,nt),yl_2D_control,'r');
```

```
xlabel('t [s]');
ylabel('y(L,t) [m]');
title('End-point deflection y(L,t) of the beam system');
legend('without control','with model-based control');

figure(9);
plot(linspace(0,tmax,nt),zl_2D_free,'b');
hold on
plot(linspace(0,tmax,nt),zl_2D_control,'r');
xlabel('t [s]');
ylabel('z(L,t) [m]');
title('End-point deflection z(L,t) of the beam system');
legend('without control','with model-based control');

figure(10);
subplot(311)
plot(linspace(0,tmax,Ttr1),ux_2D);
xlabel('Time [s]');
ylabel('u_x(t) [N]');
title('Control input u_x(t) of model-based boundary control');
subplot(312)
plot(linspace(0,tmax,Ttr1),uy_2D);
xlabel('Time [s]');
ylabel('u_y(t) [N]');
title('Control input u_y(t) of model-based boundary control');
subplot(313)
plot(linspace(0,tmax,Ttr1),uz_2D);
xlabel('Time [s]');
ylabel('u_z(t) [N]');
title('Control input u_z(t) of model-based boundary control');


function y = sgn1(x)
if x < 0
    y1=-1;
else if x == 0
    y1=0;
else
    y1=1;
end
y=y1;
```

References

1. Q.C. Nguyen, K.-S. Hong, Simultaneous control of longitudinal and transverse vibrations of an axially moving string with velocity tracking. J. Sound Vib. **331**(13), 3006–3019 (2012)
2. W. He, S. Zhang, S.S. Ge, Adaptive boundary control of a nonlinear flexible string system. IEEE Trans. Control Syst. Technol. **22**(3), 1088–1093 (2014)
3. S.S. Ge, W. He, B.V.E. How, Y.S. Choo, Boundary control of a coupled nonlinear flexible marine riser. IEEE Trans. Control Syst. Technol. **18**(5), 1080–1091 (2010)
4. T.L. Ngugen, K.D. Do, J. Pan, Boundary control of two-dimensional marine risers with bendign coupling. J. Sound Vib. **332**, 3605–3622 (2013)
5. T.L. Nguyen, K.D. Do, J. Pan, Boundary control of coupled nonlinear three dimensional marine risers. J. Mar. Sci. Appl. **12**(1), 72–88 (2013)

Chapter 12
Conclusion

The book has been dedicated to the modeling and control design of flexible beam systems. The book is divided into 12 chapters, and a variety of control methods are designed for the flexible beam systems under different conditions. The effectiveness of each control method is verified by theoretical analysis and MATLAB numerical simulation. The results of the research work conducted in this book are summarized in each chapter, and the contributions made are reviewed. The chapters of this book are interconnected and independent, and readers can choose to learn according to their own needs.

For the flexible beam system, the Hamilton principle is used to model the system. The modeling method based on the law of conservation of energy is introduced in Chap. 2. This model is composed of a governing equation (a four order partial differential equation) and a set of boundary conditions (four ordinary differential equations), to describe the flexible beam system. The models used in the subsequent chapters are based on this model and consider different situations to design different control methods. It is important to point out that the models used in Chaps. 10 and 11 are different from the model in the preceding chapters. In Chap. 10, the model is discretized by using the assumed mode method, and obtain the N-mode discrete dynamic model. In Chap. 11, the one-dimensional model is extended to a three-dimensional model, which makes the flexible beam system more closely to the actual environment.

Based on the dynamic model obtained and considering a variety of conditions, various control methods are designed in this book. The control design is mainly based on the Lyapunov's direct method, and the Lyapunov's stability theory and lemmas used in this book are introduced in Chap. 2. The main control methods involved in this book are the boundary control, the distributed control, the iterative learning control, and the neural network control. Boundary control can be derived from a Lyapunov function which is relevant to the mechanical energy based on the dynamics

W. He and J. Liu, *Active Vibration Control and Stability Analysis of Flexible Beam Systems*, https://doi.org/10.1007/978-981-10-7539-1_12

of the system. From Chaps. 3 to 7, we design various boundary control laws to solve the problems of parameter uncertainty, output constraint, input constraint (input saturation, input dead-zone, input backlash), respectively. In the control design, we combine boundary control, robust control, adaptive control and disturbance observer, and use barrier Lyapunov function, hyperbolic tangent function, input linearization, inverse backlash dynamics, etc. The distributed control method is a traditional control method, which is easy to understand and design. However, it is difficult to implement because of the need for more sensors and actuators. The iterative learning control is employed without perfect knowledge of the target system, and possesses relative simple structure and almost model-free nature. By learning from the past control experience, system performances are improved. The neural network control requires relatively less information of the system dynamics. It has been proven that artificial neural networks are able to approximate a wide range of nonlinear functions to any desired degree of accuracy under certain conditions.

This book systematically introduces the basic theory, method and application technology of dynamic modeling and control design for the flexible beam system. It is the crystallization of many years' teaching and research work carried out by the authors. It also integrates the new achievements made by domestic and foreign counterparts in recent years. On the other hand, there are still many problems that can be studied for the flexible beam system.

(1) The further study for the three-dimensional flexible beam:

In this book, we introduce the basic boundary control method of a three-dimensional flexible beam. It is worth mentioning that other control methods like neural network control, iterative learning control, and distributed control also can be used in the three-dimensional flexible beam system. In addition, the control of the three-dimensional flexible beam with input constraint (input saturation, input dead-zone, input backlash) and output constraint is also a meaningful research direction.

(2) The fault tolerant control:

The study of fault tolerant control is the problem of control when the system fails. Passive fault-tolerant control is a scheme similar to robust control technology. It adopts a fixed controller to ensure that the closed-loop system is insensitive to specific faults and keeps the system stable. The active fault tolerant control is to redesign a control system according to the desired characteristics and make the whole fault system stable after the failure.

(3) The uncertainty and disturbance estimator (UDE):

The robust control method has been used in this book. In recent years, another method to handle the uncertainties and disturbances which is called uncertainty and disturbance estimator (UDE)-based control, has received more and more attention. UDE compensates the disturbance through an estimation by filtering the system input and state information. Due to the simple structure and convenient implementation in real time, UDE has been applied to many linear and nonlinear systems.

(4) The experiment for the flexible beam system:

The verification of control effect in this book is based on theoretical analysis and numerical simulation. In the future, the platform of the flexible beam system will be set up. It will be more convincing to verify the effectiveness of the control method through experiments.

MIX
Papier aus verantwortungsvollen Quellen
Paper from responsible sources
FSC® C105338

If you have any concerns about our products, you can contact us on
ProductSafety@springernature.com

In case Publisher is established outside the EU, the EU authorized representative is:
Springer Nature Customer Service Center GmbH
Europaplatz 3, 69115 Heidelberg, Germany

Printed by Libri Plureos GmbH
in Hamburg, Germany